Lincoln Oldshue

Urino-Pathology

The Uroscopian System of Diagnosing Diseases

Lincoln Oldshue

Urino-Pathology
The Uroscopian System of Diagnosing Diseases

ISBN/EAN: 9783337715090

Printed in Europe, USA, Canada, Australia, Japan

Cover: Foto ©berggeist007 / pixelio.de

More available books at **www.hansebooks.com**

URINO-PATHOLOGY;

OR

THE UROSCOPIAN SYSTEM

OF

DIAGNOSING DISEASES,

BY

OCULAR INSPECTION, CHEMICAL ANALYSIS, AND MICROSCOPIC EXAMINATION OF THE URINE;

ACCOMPANIED BY

AN ILLUSTRATIVE CHART

OF URINARY DEPOSITS,

REPRESENTING THE MICROSCOPIC APPEARANCE OF MORE THAN SIXTY DIFFERENT VARIETIES OF INGREDIENTS FOUND IN THE URINE.

By L. OLDSHUE, M. D.,

PROFESSOR OF PATHOLOGY IN THE ECLECTIC MEDICAL COLLEGE OF PHILADELPHIA, AND MEMBER OF THE AMERICAN ECLECTIC MEDICAL SOCIETY OF PHILADELPHIA.

PITTSBURGH:

PRINTED BY A. A. ANDERSON & SONS, Nos. 67 AND 69 FIFTH STREET, DISPATCH BUILDINGS.

1864.

RECOMMENDATION.

FROM time immemorial, physicians have examined with more or less attention the renal secretion, with the view of better understanding morbid conditions of the human system. Very general directions, with only here and there a few particular cases, have slowly crept into our medical text-books, until very recently.

Of late the attention of the medical world has been called to the subject in two different ways, to-wit:

1st. Practically, by the practice of what is called *Uroscopia*.

Uromantia, or the art of diagnosing diseases by the simple inspection of the urine, has not been practiced exclusively in this country; still, many of the most learned in the Profession have paid considerable attention to the subject.

In the Germanic States, it is affirmed by some, to be better understood. In this way our empirical knowledge was of a practical nature, which scientific analyses must render more definite and intelligible.

2d. Valuable scientific treatises have lately appeared, demonstrating, beyond a doubt, that even empirical claims fall far short of facts.

Now, by the aid of these works, united with the experience of practical Uroscopians, a very important addition to our present stock of knowledge, might be embraced in a scientific work on Uroscopia. Its importance and practicability has been ably set forth in a series of articles from the pen of DR. OLDSHUE, of Pittsburgh. He is presenting to the reader the Uroscopian System, as he was taught, and as he has practiced it, though not exclusively, but as an important auxiliary, to correct diagnosis. It is a subject which invites investigation, and which promises valuable practical reward for careful researches. DR. OLDSHUE has presented us with a mass of testimony attesting the value and importance of understanding this subject, quoting the opinions of those who rank first as medical investigators.

We can not here omit noticing an incidental remark of Prof. Buchanan, relating to the intimacy sustained between all the secretions, and especially the *renal*, and the condition of the brain. It is presumed that he understands the subject of physiology, not only, but is familiar with the opinions of physiologists. He says:

"It is regarded by physiologists as an established truth, that the action of the brain, as well as every other organ of the constitution, is accompanied by chemical changes in its substance, consisting of the disintegration of the nervous matter, and the nutritive deposit of the freshly-assimilated particles of the blood. It has been ascertained that these changes are accompanied by corresponding changes in the secretions, and that the kidneys, of all organs, appear to sustain the most intimate relation to the brain."

Put these remarks (says Prof. O. Davis) beside the clinical statistics furnished by the General Hospital, at Vienna, according to the testimony of Prof. White, and we see this "established truth" is so considered by those for whose opinions we entertain respect.

Dr. Oldshue is doing a valuable work, and we hope to receive many more articles from his pen on the Uroscopian System.

Rochester, December, 1851.

OPINIONS OF THE PRESS.

"Dr. Oldshue is a graduate of the Reformed Medical College, and has sustained an excellent character as a man, and is devotedly attached to his profession, and attentive to his business."—*M'Keesport Standard*, 1856.

"Dr. Oldshue, a number of years since, practiced Medicine in this county, and is well and favorably known to many of our citizens. He has acquired considerable notoriety for his success in treating diseases."—*Connellsville Enterprise, Fayette Co.*

"He has acquired a large reputation for his successful treatment of diseases, and commands an extensive practice."—*Wellsville Patriot, O.*

"He has become a perfect master in his profession, and by a careful study of the Uroscopian System, has, by a series of experiments and years of practice, acquired such a ready knowledge as to enable him to detect the most complicated case, by an examination of the *urine*."—*Greensburg Herald, Westmoreland Co.*

"He is a regularly educated Physician of over fifteen years practice, ten years of which time he has been in Pittsburgh, where he is well and favorably known, both by the people and the Profession, and, as we learn from the patients themselves, has been very successful."—*Independent Banner, Clarion Co.*, 1856.

"He has acquired a well deserved reputation for the cure of the worst maladies known to the human system. His treatment is both safe and effectual, and has produced the most astounding results. We have this from some of the most respectable persons in our county, who have sent to him from this place."—*Waynesburg Eagle, Greene Co.*

"We are well acquainted with the treatment adopted by Doctor Oldshue, as well as the successful results which have so uniformly attended his professional labors."—*Indiana Independent, Indiana Co.*

"He treats all kinds of diseases. His study and knowledge of the Uroscopian System, enables him to treat cases from a distance, by an examination of their *urine*. He has in this respect been more successful than any Physician within our knowledge."—*Brookville Jeffersonian.*

"In Medical Practice, as in everything else, the best test of a Theory is its practical results. The success of Dr. Oldshue, during a Practice of ten years in Pittsburgh alone, affords sufficient evidence of professional skill and ability."—*Blairsville American, Indiana Co.*, 1856.

"Fifteen years practice, assisted by close study, as is the course of the Doctor, places him in no subordinate position in the line of his profession." —*Wellsville Patriot.*

"He is a gentleman in every respect qualified for his position. His career in the line of his profession has been a successful one, and he has seldom failed to both relieve and cure his patients."—*Connellsville Enterprise.*

"An improvement has been effected in the treatment of diseases under a Progressive and Reformed System of Medical Science; and Dr. O. has added more perhaps to this reform than any man in the West. He treats all kinds of diseases; and his success has been almost universally attended with the most satisfactory results."—*Columbian Republican, O.*

"He is not unknown to many of our citizens, numbers having visited him from this place for professional relief. Testimonials from different parts of our county, attest his success in curing different diseases of a difficult and complicated character."—*Clarion Democrat, Clarion Co.*

"He is well acquainted with all the late improvements in the treatment of disease; and also understands the Uroscopian System, and those from a distance, who can not attend him personally (a vial of *urine* being sent him), will receive a thorough examination, and medicines can be sent them."—*Armstrong Democrat, Armstrong Co.*

"He treats all kinds of diseases common to the human system, and is thoroughly studied in all the modes of practice. The writer of this, within the past year, has visited the principal Medical Institutions in Philadelphia and Cincinnati, and uniformly heard but one opinion expressed of the gentleman in question (viz: Dr. Lincoln Oldshue), and that was, of unqualified approbation of his great medical research, as well as his honest and honorable mode of Practice."—*From the Pittsburgh Union.*

"Dr. L. Oldshue, who for the last ten years has been located in Pittsburgh, formerly practiced Medicine in our county, and is well known to many of our citizens. He is a regular graduate in his profession, and the best evidence of his SKILL is his acknowledged success, which as we understand, from the patients themselves, that have visited him from this place, has been remarkably good."—*Genius of Liberty, Fayette Co.*

"He is well and favorobly known to most of our readers, and is a gentleman of talent, experience and ability, and a successful physician in all he undertakes."—*Washington Review, Washington Co.*

"He has a good practice, and is worthy of all confidence. The better he is known the more he will be esteemed and employed."—*Pittsburgh Evening Chronicle.*

WORTHY OPINIONS.

"The objection often urged against the possibility of a minute acquaintance with urinary deposits being available in practice, no longer exists; a minute or two being sufficient for the observer to learn the nature of any variety of sediment."
GOLDING BIRD, A. M., M. D.

"We can arrive at a more accurate knowledge respecting the nature of diseases from examining the urine, than from any other symptom."
DR. BRAITHWAITE—*Retrospect.*

"Whatever may be the disease, the urine seldom fails in furnishing a clue to the principles upon which it is to be treated."—*Eberle's Therapeutics.*

"The different conditions of the urinary discharge, seems to indicate a corresponding difference in the constitutional affection, to which they belong; and we entertain hopes, that hereafter, and under a more accumulated experience, they may be found important guides in practice."
JOHN BLACKALL, M. D.

"The importance and practicability of a scientific work on Uroscopia, has been ably set forth in a series of articles from the pen of Dr. Oldshue, of Pittsburgh. He is presenting to the reader the "Uroscopian System" as he was taught and has practiced it, as an important auxiliary to diagnosis."
PROF. O. DAVIS.

"The great impulse that has lately been given to the study of Chemical Pathology has thrown considerable light on many obscure subjects. The Urinary Secretion is the chief, and at the same time the most important subject of attention in this branch of science."—*Markwick on the Urine.*

"From the physical and chemical state of the urine, the attentive, observing physician, may obtain a great quantity of information for ascertaining and establishing a diagnosis. More than all other signs, the correct examination of the sediment is of importance to the physician."
DR. F. SIMON.

"The examination of the Urine in disease is now regarded as one of the most important aids in diagnosis, and which it would be alike injurious to the welfare of the patient, as to the credit of the practitioner to avoid."
DR. G. BIRD.

"Dr. Oldshue is presenting us with a mass of testimony, attesting the value and importance of understanding this subject, and doing a valuable work. We hope to receive many more articles from his pen on the "Uroscopian System."—*Eclectic Journal of Medicine*

"To argue that such investigations are idle, are as absurd as unfortunately they are frequent. But there is, however, one consolation in this matter, which is, that those who are most ready to urge this view, and to decry its utility, are such as are least acquainted with its detail."
J. W. GRIFFITH, M. D., F. R. S.

INTRODUCTION.

Although the various diseases which are prevalent in our country appear to be sufficiently defined by our medical authors, yet no observing practitioner can have had much experience at the bedside of the sick, without frequently finding it peculiarly difficult, and ofttimes impossible to ascertain the precise nature, extent, and locality, of the diseases he is required to combat.

These difficulties are acknowledged by all medical authors, and experienced practitioners, and it has been the opinion of some of our most able medical men, that a majority of deaths which occur when a physician has been timely employed, are occasioned by their mistaking the nature of the disease they are required to cure. And, should infancy, insanity, idiocy, delirium, hypochondria, paralysis, insensibility, or any other cause, prevent the patient from describing the symptoms, the difficulty of determining its nature is very much increased.

With a knowledge of these facts, we, therefore, through the whole course of our study and practice, in addition to the regular means, have paid particular attention to that branch of medical science called *Uroscopia*, or the detection of diseases by an examination of the urine, our medical duties being constantly brought under that mode of distinguishing diseases.

To make pretensions to a new or mysterious system of diagnosing diseases, is not the object of this work, but to direct the attention to a too much-neglected auxiliary mode; believing that a better acquaintance with its principles will assist in arriving at a more intimate knowledge of the nature of the different diseases that afflict the human family, and thereby enable us to calculate a more positive method of treatment.

The Uroscopian system is not of recent origin, or at least, the practice of analyzing and inspecting the urine, for the better discrimination of diseases, has long been conducted by many medical men. The march of improvement, however, has from time to time, developed fact upon fact, relative to the urinary secretion in the different diseases, which development has given such impetus of late, to this growingly, interesting subject, that we have attempted to reduce these facts to a more decided form.

This has been a labor of no little magnitude; but the ample opportunity of investigating the subject, which it has been our lot to enjoy for the last twenty years—having conducted the examination of over three thousand specimens of urine, annually, or an aggregate of over fifty thousand cases—we feel justified in this effort to systematize the course, and hope to be able to present such facts, and exhibit such principles and rules, for the examination of the urinary secretion, that a more accurate diagnosis of nearly all forms of disease may be thereby deduced with more nearly mathematical precision than can possibly be done without. These principles and rules, when thoroughly understood, will many times conduct the practitioner through the most complicated cases, to the most minutely correct conclusions; for, as in the English language, which comprises more than sixty thousand words, all of which are distinctly divided into a few classes, or parts of speech, and are all governed by a set number of rules, whereby each particular word in the whole vocabulary is not only referred to its class, but is distinctly assigned to its own particular place in the millions of sentences which are comprised in the language, so with the principles and rules which govern the examination of urine in diseases; by these, each specimen of urine may not only be referred to the class of disease to which it belongs, but may be traced to its division or subdivision, and even assigned to its own particular place in the series.

Life is a peculiar property of organic matter, of which we know nothing, but from the phenomena it displays in an organized body. All organized bodies have a limited existence of form, however, and pass through certain ages, the terminations of which are called death; and anything which reduces these limitations, or cuts short these ages, may be termed disease—the antagonist of life.

Disease is strictly an activity. Its first appearance is the first of a series which in the nature of things is continually progressing. Therefore, so long as the antagonism of vitality and disease exist in any subject, there will be antagonistic series going on. Each disease is a series of itself, and each particular form of disease, one only of its series, these series being of every gradation from the slightest predisposing influence, to the complete destructien of the age of the existence, or life.

This antagonist of life manifests its attack upon the body first, in the blood, then upon other parts, in a variety of ways. And in whatever way, the urine soon takes on the characteristic indications either in optical appearances, chemical composition, or microscopic characters, or all of these together.

To facilitate the labor of the study of its microscopic character, we have prepared a Chart to accompany the work, arranged in the most perfectly scientific order; its divisions, lettering and artistic workmanshih, being the most complete, systematic and convenient, of any now in use, as a single glance at it will show.

For the facts in elucidating the "general principles," as also in determining special diseases and causes, we have drawn largely upon the researches and labors of Bird, Reese, Markwick, Bowman, Bennet, Andral, Prout, Griffith, Beach, Paine, Newton, Simon, Williams, Wilson, Taylor, Bartlett, Wood, and a number of others, wherever their observations accorded with our own.

The urine being composed of such parts of the human system as have been wornout or diseased, and can no longer remain therein, without detriment to its healthy operations, together with the super-

2

fluous matter collected from every part of the system and transmitted to the blood, and separated therefrom by the kidneys ("the grand drain established by nature for purifying the system of superfluous or unhealthy matter"), it is rational to examine that secretion in disease to ascertain the healthy and unhealthy operations that are thus going on.

Again, if any part of the system becomes diseased, its vitality, of course, is lessened, and consequently that part has less power to resist the tendency to decomposition; hence, there will be more of this diseased substance or wornout particles, passed from this diseased part, than from the healthy parts of the system, at the same time. And, as the different parts of the system are of different composition, colors, consistency, etc., so much so that almost any person can distinguish merely by their appearance, the lungs from that of the heart, liver, stomach, kidneys, brains, etc., or one of them from another, so the refuse fluid, or wornout or *diseased* particles of these parts, are also different one from another; therefore, it is plain that, if any organ gives out more than its customary share, it will alter the general properties and composition of that fluid whereby it is carried off.

Now if these facts be true (and for their confirmation we appeal to our medical authors, and such of our medical men as understand the pathology of diseases), every different disorder will impart its own peculiar stamp to the urine, and we need only know what that peculiar stamp is, to pronounce upon the nature, extent, and character of the disorder, with almost undeviating certainty.

Nor is this mere idle theory; for although the subject has attracted but little attention for years, and has been almost neglected in general practice, yet all our most celebrated authors, Cullen, Good, Gregory, Dewees, Dunglinson, Eberle, Beach, etc. etc., unite in ascribing to almost every disease of which they treat, a characteristic appearance of the urine. And hence, when we consult medical authors the above opinion is confirmed.

And when we consider the structure of the human system, the composition and circulation of the blood, the peculiar chemical changes in that fluid in disease; and the nature, composition and origin of the urine, with the ahemical changes and different appearances, consequent upon the changes in the blood, we are induced to pronounce unhesitatingly, that, it is a valuable test for distinguishing disease. And yet it has long been a matter of *wonder* to some medical men, how anything like a correct diagnosis of any disease can be made from the inspection or examination of the urine. This wonder ceases, however, as they become more practically acquainted with all the varied chemical qualities and characteristic appearances of that fluid as voided in the different classes and varieties of disease.

To become so acquainted has hitherto been a matter of very difficult accomplishment, in consequence of the confusion through which the system has been laboring, for want of being reduced to a systematic form. The very limited knowledge of it, as taught in our medical schools, seldom renders the student capable of applying it, beyond a mere auxiliary in urinary diseases, that being the extent to which observations are generally made on this subject.

We claim, however, to have made further investigations upon this subject, which investigations have demonstrated to us the facts, that every *class* of disease communicates its separate and distinct characteristic appearance; and every *particular* disease, its own peculiar stamp to the urine; and that these investigations and the discovery of these facts, have enabled us to lay down certain rules for the examination of that secretion, whereby these peculiar qualities and characteristics may be almost invariably detected; and thereby to determine not only the nature of the disease, but even to form a very correct opinion as to its extent also, from a mere inspection of a specimen of urine voided at the time.

We claim more: we claim to have so arranged this system, as to be able to teach it to students with as much success and general satisfaction, as is given in surgery, materia medica, or the practice of medicine in general.

For the convenience of such teaching, as well as for the more correct application of the principles to general practice, we have reduced the system to a prescribed form, regulated according to certain general rules.

First, then, is the classification. For, as diseases are very readily divided into classes, each class having its own peculiar characteristics, whereby it is readily distinguished from others, so is the urine *naturally* classified, as separately and distinctly as are diseases themselves. Each different class of disease, as before observed, communicates its characteristic trait to the urine, and each particular variety of disease imparts its peculiar stamp to the same; hence we will at once observe, that by first classifying the urine according to these characteristics, then particularizing according to these peculiar stamps, we will have no difficulty in arriving at the precise nature of the malady.

By this arrangement we are enabled, almost at first sight of a specimen of urine, in nearly every instance, to determine the class to which any such specimen, and consequently, the disease, belongs—whether inflammatory, febrile, eruptive, dropsical, nervous, etc., even without asking a single question concerning the patient.

When the class is thus determined, a note of the same is taken down, and a search instituted for the *peculiar stamp*, whereby the variety of the class is characterized. If we find ourselves unable to detect, by ocular inspection, the characteristics indicating the more particular disease, the specimen is then submitted to the most convenient appropriate chemical test, in the selection of which test we are governed, and generally very correctly too, by the class also, this being the index to the farther investigations, which enable us to determine more precisely the malady.

When this is done, the specimen of urine is carefully sealed up, and preserved for re-examinations by our students; and so we continue to collect and preserve all the important specimens, until a cabinet of several hundred of every grade and variety is gathered together.

Each specimen is labelled according to its class, and also numbered according to its peculiarity or variety, such numbers being severally noted down in a book kept expressly for the purpose, which notes serve as a guide to the students, in the numberless re-examinations and comparisons of specimens one with another, which are necessary to a perfect knowledge of the science. For, like the practiced ear to music, which is enabled by the constant exercise of the *sense* of hearing to detect the most minute discord in the most intricate piece; so, by the oft-repeated, or continual exercise of the *faculties* upon this subject, they are rendered so acute as to detect the difference in the most difficult and complicated cases.

These daily re-examinations and comparisons of several hundred specimens, together with the daily accessions of new cases, which are constantly being brought to a practical Uroscopian of reputation, can not fail to teach him *something* that is not to be learned by a few observations alone.

Hence, having observed that it requires a greater number of examinations than generally comes under the notice of medical men, who make no pretensions to a knowledge of this branch of the profession, to enable them to come into full possession of a thorough knowledge of this science, we will readily see the advantages of our plan and course of study.

The student must persevere beyond a few experiments alone, and must look through premises to conclusions, and must conduct his experiments and investigations through the whole classification, before he will be enabled to draw any pathological deductions. When he is able to classify the urine in accordance with the disease, he then has hold of the key, as it were, which unlocks to him the door to farther investigations, by persevering in which, the Uroscopian system will be developed to his pleasing satisfaction.

The system must be thoroughly understood before it can be practiced at all. No one can succeed without. An attempt of anyone to palm himself upon the community as a practical Uroscopian, without a thorough knowledge of its principles, would discover him perhaps, in the very first case presented for examination. His failure to at once detect and describe the disease, would exhibit his ignorance, and drive him from his false position.

It would be impossible for any person, however shrewd a Yankee, or apt in the "art of guessing," he might be, to pronounce at once the precise nature of the disease, in a sufficient number of cases to acquire a reputation. And, a single failure in any case, would almost be utter ruin, so incredulous are the community on this subject. Besides, the patients themselves in many cases are good judges of their own aches and pains, and can tell at once whether their symptoms have been described or not. And on the other hand, if they hear proclaimed with certainty the organ or organs affected, and the true nature of the disease and all the attending symptoms fully described, they see at once the positive application of the system and are struck with astonishment at its correctness; and hence it is, in part, that a good business is always secured by a practitioner of this class. Not only has he the advantage of a more complete knowledge of the disease, but a clue at the same time is given to the treatment thereof; and thus a twofold advantage is derived: the patient in his chances of recovery, and the practitioner in point of success.

To show that we are not claiming too much for our system when we propose to make it an auxiliary of great value, in diagnosing many diseases, we will exhibit the opinions of several medical authors who rank high in the Profession, Eberle, Prout, Braithwaite, Golding Bird, and Simon.

Prof. Eberle, says, "The urine being secreted from the blood during morbid vascular excitement, the substance with which the blood is surcharged, together with the substances cast from the excited organs will be imparted to the urine. The appearances and character of the urine often afford valuable diagnostic indications. We find, too, that the solution of diseases is more frequently attended by a critical discharge of urine than by any other of the excretions."

Dr. Prout, says: "The appearances of the urine were at one time regarded as of the utmost consequence in forming a proper opinion of the character of diseases. At present, this secretion is too much neglected; by an attention to it, we will often be greatly aided in our judgment of the nature of the diseases. Generally speaking, nothing can be more opposite than the conditions of the system, and consequently, the principles of practice indicated by a diminished or increased flow of urine. Hence, they are symptoms of primary importance in all diseases in which the urine is concerned; *and whatever may be the disease*, seldom fail of furnishing us with a clue to the principles upon which it is to be treated." (Vol. 2, Eberle's Therapeutics).

Prof. Braithwaite says: "We can arrive at a more accurate knowledge respecting the generality of diseases from examining the urine, than from any other symptom."

Golding Bird, speaking of the morbid conditions of the urinary secretion, says that we are to regard them as one of a series of pathological changes going on in the system, and more valuable than others as an index of disease, in consequence of the facility with which it is detected; that the urine pumps off any excess of fluid that enters the circulation and causes disease.

Dr. Simon says: "There is not a more important sign offered in disease, than in the urinary secretion."

Hundreds of others might be referred to, bearing equally valuable testimony in its favor; many of whom will be quoted in the body of this work, whose opinions will add weight to the truths it contains, and assist in convincing many of the facts that, of all the means of determining the precise nature and extent of diseases, none other is so universally to be depended upon, and none so easily and satisfactorily accomplished.

ERRATA.

On page	31	line	24, for E. C. F., read E. and F.
"	47	"	10, for rapidity, read respiration.
"	67	"	5, for in blood, read of the blood.
"	87	"	21, for exertion, read excretion.
"	137	"	19, for decreased, read diseased.
"	153	"	13, for from two, read from one to two.
"	156	"	20, for caeries, read caries.
"	176	"	4, for Arnold, read Andral.

URINO-PATHOLOGY;

OR,

THE UROSCOPIAN SYSTEM

OF

DIAGNOSING DISEASES.

SIMON in his "Chemistry of Man" divides the proximate constituents of the animal body into two great classes, the MINERAL and ORGANIC. The *Mineral Constituents* he classes in three groups comprising 1st, those which are of service in the animal body in consequence of their physical properties; 2d, those which effect important objects in the system by their chemical actions; and 3d, those which, being only incidentally present, may be eliminated without exerting any unfavorable effect on the economy."

In the first group, and first in the importance of its physical properties to the animal organism is ranked *Water.*

Second in importance to water, is ranked *Phosphate of Lime*, or *bone-earth*, as occurring in bone, blood, milk, urine and fæces.

Third in rank is placed *Carbonate of Lime*, as forming the principal part of the skeleton.

Fourth, *Phosphate of Magnesia*, as being frequently associated with phosphate of lime, and occurrng in bone, blood, milk, urine, fæces, concretions, etc.

Phosphate of Magnesia and *Ammonia*, or as it is frequently termed, *Ammoniaco-Magnesian Phosphates*, is given as a distinct salt, which is present in certain states of the urine.

And last, *Fluoride of Calcium*, as occurring in the animal organism in minute quantity.

In the second group, the first importance is given to *Hydrochloric Acid*, as being useful by its *chemical* properties.

The second in importance is *Hydrofluoric Acid.*

Third, *Chloride of Sodium.*

Fourth, *Carbonate of Soda.*

Fifth, *Phosphate of Soda.*

Sixth, *Chloride of Calcium.*

Seventh, *Chloride of Iron.*

Eighth, *Iron*, as being the principal coloring matter of the blood.

In group third, the incidental constituents are placed in the following order:

First, *Chloride of Potassium.*

Second, *Alkaline Sulphates.*

Third, *Carbonate of Magnesia.*

Fourth, *Manganese.*

Fifth, *Silica.*

Sixth, *Allumina.*

Seventh, *Arsenic.*

Eighth, *Copper.*

Ninth, *Lead.*

Tenth, *Ammoniacal Salts.*

While each of these have been considered as proximate constituents of some part of the animal body, nearly all have been found in the urine, either as useful, or an incidental constituent of that fluid.

The Organic constituents are arranged in two principal groups—the *Nitrogenous* and the *Non-Nitrogenous* matters.

First in importance of the Nitrogenous constituents, is placed *Protein.*

Under the head of *Protein*, three very important compounds are considered; compounds which are found to constitute the greater part of the animal body. These are Albumen, Fibrin, and Casein.

The name *Protein*, which signifies "I am first," is regarded as the starting-point of all the tissues. And it is said that

Protein, in every respect identical with that which forms the basis of the three aforesaid animal principles, may be obtained from similar elements in the vegetable kingdom.

There is *Vegetable Albumen*, *Vegetable Fibrin* and *Vegetable Casein*, and "the chemical analysis of these three substances has led to the very interesting result that they contain the same organic elements, united in the same proportion of weight; and what is still more remarkable, that they are identical in composition with the chief constituents of blood, to-wit: animal fibrin and albumen."

Second in importance ranks *Albumen*, which is but a modification of protein, and occurs in large quantity in all the animal fluids that contribute to the nutrition of the organism.

Third in importance ranks *Fibrin*, which is but a modification of the first, also.

Fourth is *Casein*, which substance constitutes the most important ingredient in the milk of the mammalia.

Fifth in order is *Pepsin*, a substance which constitutes the most essential portion of the gastric juice.

Sixth, *Ptyalin*, a peculiar animal matter that exists in the saliva.

Seventh, *Gelatin*. This does not exist as *gelatin* in the animal tissues, but may be formed from them by the action of boiling water.

Eighth, *Pyin*, a peculiar substance which occurs in pus.

Ninth, *Extractive Matters*, an amorphous mass of organic nitrogenous matter which remains after the protein compounds and the salts have been removed from the animal fluids.

Tenth, the *Coloring Matters* of the blood, bile and urine, to wit: *Hematin*, *Biliphin*, and *Uroerthyrin*.

Eleventh, *Bilin*, the most important constituent of bile.

Twelfth, *Urea*, the principal constituent of the solid residue of normal human urine. "It is found in considerable quantities in the blood after extirpation of the kidneys, also in certain pathological conditions in which the renal functions are not properly discharged."

Thirteenth, *Uric Acid*, a constituent of the urinary secretion in apparently all classes of animals.

Fourteenth, *Hippuric Acid.* This is said to sometimes occur as an ingredient in healthy human urine. It is readily obtained, however, from the urine of the horse or cow.

Fifteenth, *Uric Oxide.* This is said to be an ingredient in severe cases of vesical calculi.

Sixteenth, *Cystin.* This is an occasional constituent of the urinary secretion, and is sometimes found as a crystalline deposit, especially in scrofulous and cachetic diseases.

These are the *Organic Nitrogenous* constituents of the animal body numbered in the order of their importance.

Next comes the *Non-Nitrogenous*, and the most important among these stands—

First, *Animal Sugars.* Sugar of milk is an important constituent of the milk of the mammalia.

Second, *Fats.* Under this head is included "the various *Non-Nitrogenous* compounds, which are insoluble in water.

Third, *Acids.* Under this head is included *Lactic*, *Oxalic* and *Acetic* Acids.

Lactic Acid is regarded by most chemists as a constituent of almost all the fluids of the animal body.

Oxalic Acid is not considered one of the normal constituents of the animal organism, but is a very common ingredient of morbid urine when combined with lime, and in the form of *Oxalate of Lime*, is found in the urine in dyspepsia, etc.

Acetic Acid has been found by Tiedeman and Gmelin in the gastric juice, and is asserted by some chemists to be a constituent of urine.

These are the proximate constituents of the animal body—Mineral and Organic; and while the study of the variations of these would give most important information, the limits of this work make it expedient for us to confine ourselves to this mere relation of them at present. It will be observed that a large number of these perfectly distinct substances enter into the composition of the blood and urine. And as the examination of the *variations* in the urine, in relation to pathological conditions, is the object of our labors, we will be confined more particularly to these.

EXAMINATION OF URINE.

THERE are three prominent modes whereby the urine may be profitably examined for the better discrimination and more correct diagnosis of many diseases.

First—OPTICALLY only. Noticing its transparency, opacity, fluidity, turbidity or limpidity: Whether it be colorless, amber, saffron, red, yellow, brown, olive, green, blue, black, or tinged only with either; whether pale, bright, deep, scarlet, crimson, dingy, or blood-red color; whether straw, orange, ash, icteritious, or clay-colored; whether chesnut, reddish, yellowish, greenish, bluish, or blackish-brown color; whether pale, deep, dark, apple or sea-*green* color. And so of every shade of color and of all mixtures of colors, from the milk-white, down to the muddy, dirty-water-like, blackish-brown urine of putrefaction.

In this examination you will also observe whether it be clear, cloudy, ropy, thick or muddy.

If cloudy, whether it coagulates in a sheet or table in the center of the vessel, rises to the top, or falls to the bottom of the urinal; whether it throws up a cream-like pellicle to the surface, deposits a sediment at the bottom, or crystalizes on the sides of the vessel.

All these have their diagnostic importance, and each expresses a significant fact of great value, in the progress of examinations, and many times, alone throw a flood of light upon the nature of the case.

The principal coloring matters of the urine are, however, generally few, and of very easy determination. Where but one coloring matter is present there will be but little difficulty to decide, but in the combinations which so variously occur, we may sometimes be considerably perplexed. We will here notice some of the most common, with their peculiarities and qualities.

Purpurine, is a peculiar coloring matter which is always combined with urate of ammonia, and varies in tint from a mere flesh color to the deepest carmine. It was thought by some to be identical with murexid or purpurate of ammonia. "It merits considerable notice," says Bird, "from the serious lesions its presence frequently indicates." "When present in urine, it will often communicate to it so intense a color, as to cause the patient to report his urine to be bloody." It indicates hepatic derangement, affection of liver, spleen, etc.

Cyanourine, gives a blue color to urine, and deposits in a blue powder. Diluted acids dissolve it, however, and change it to brown or red, according to the proportion of acid present.

Indigo, which is sometimes found in urine, gives it a dark-blue color.

Hemaphaen, is the yellow coloring matter of urine, "and it is probably," says Bird, "to its presence in excess, that the jaundiced hue of persons in a state of *anemia* and chlorosis is owing." "The normal amber color of urine is probably owing to a mixture of this pigment with a red one."

Cholestrine, is supposed to give the coloring principle to bile. It assumes a blood-red color from the action of sulphuric acid. It frequently occurs in a crystalline state in animal fluids, in the blood, bile, fluid of hydrocele and uterine hydatids. It forms the chief ingredient of biliary calculi. It often gives the urine a deep-brown color.

Hœmatosine, is the coloring matter of the blood, and is the most general coloring matter found in the urine also, the urine being secreted entirely from the blood, must partake somewhat of its coloring matter very generally.

This coloring matter varies in urine, from the brightest red, to the darkest brown, according to its combination with other substances, and to its excess or deficiency therein.

Hematoxylin, administered as a medicine, will often, by the red color it communicates to the urine, lead to an unfounded suspicion of the existence of hematosine.

Mucus, pus, milk, fat, etc., all have the effect of changing the color and optical appearance of the urine, but may be known by their different qualities.

"The knowledge of urinary disorders," says Dr. Aldridge, "has so rapidly progressed within the last few years, as to have now arrived at least to a *level* with other departments of medical science, but unfortunately this knowledge is very little diffused among the members of the medical profession."

As a guide to the learner we would recommend specimens of urine to be collected from patients laboring under the different diseases, and in fact all different stages of the same disease; carefully preserve the same, not only for comparison one with another, but for comparison with new specimens that may from time to time be added.

A cabinet of several hundred of these specimens, comprising as they might do, urine from every grade and variety of disease, would be of valuable asssistance to the learner, and even to the old practitioner it would not be without its importance in making up his diagnosis in many cases.

Those who would obtain a knowledge of this important branch of the medical art, by pursuing such a course, keeping in view the *age* and sex of the patients, having a knowledge of the appearance and standard properties and qualities of healthy urine, as also the pathological conditions of the system, may learn many valuable lessons thereby.

And when to these lessons are added the knowledge to be derived from a chemical analysis of the urine in the different diseases also, and finally, that, from a microscopic examination of the same, the precise nature of almost every malady may be determined with almost mathematical precision.

Second—CHEMICALLY. Which is but the aid of chemical action of re-agents, and manual assistants, to the first.

After the urine is fully surveyed and all its peculiarities as to color, consistence, etc., in the form as first discharged, is noticed, the changes which certain chemical agents may produce thereon will sometimes establish certain other facts yet unnoticed, or at least the more confirm what we have already learned.

The character of these changes are, principally determined by their optical appearances also, to be noted, however, at the time. Upon the nice distinctions of color and other conditions, which the eye only can make, depends most of our positive knowledge of anything. And what more reliable sense is there belonging to man than the sense of sight; yet it is here greatly aided by these chemical manipulations.

The first chemical manipulation involves the determination of the acidity, alkalinity, or neutrality of the urine. This is done by immersing in the specimen to be tested, red and blue litmus paper alternately. If alkaline the red color of the paper will be changed to blue. If acid the blue color will be changed to red. And if no change in color occurs the urine is neutral.

The following from the lectures of Dr. Aldridge is important to be remembered: "If the urine be very acid in its reaction, and deposits a red, pink, buff-colored, or white precipitate, the chances are that the precipitates are urates, or uric acid. If upon pouring out the urine the bottom of the vessel be stained with an even powdery coating, it is likely to be *lithates.* If hard crystalline grains adhere to the sides and bottom of the vessel, *uric acid.* If either acid, alkaline, or neutral urine be *turbid* on emission, deposits a whitish, yellowish or red sediment and is not rendered transparent by heat, it contains either phosphates, oxalate of lime, cystine, mucus, pus, or blood. If *phosphates*, it will be rendered transparent by acetic acid. If *cystine*, it will be dissolved by water of ammonia. If *oxalate of lime*, it will not be affected by acetic acid nor aqua-ammoniæ. If not dissolved by any of the above, it is mucus or pus, or if red it is blood."

These constituents of urine are all easily detected by the several chemical *tests* that will be given in other parts of this work, and when the facts thus obtained, combined with microscopic revelations, all correspond in a given case, they will be a tolerably sure guide to correct conclusions.

Third—MICROSCOPICALLY. Having examined a specimen of urine optically alone, then chemically in connection, it is ready for microscopic inspection. In fact the microscope might be constantly at hand during all the examinations, and should be applied to frequently in all the chemical changes. If by the microscopic examination of the merest fragment of a bone, a naturalist is enabled to pronounce with certainty not only the *natural family* to which such fragment belongs, but to determine with equal certainty whether it was part of a thigh or a wing-bone of a bird, as also from microscopic examination of the minutest fragment of the same bone is enabled to estimate the size of said bird, as was Prof. Owen, how much less difficult should it be for the medical practitioner to be enabled to ascertain by microscopic examination of any urinary deposit, not only the precise character of such deposit, but the nature of disease and the condition of the part whence such deposit is derived.

"During the last few years," says Carpenter, "the microscope, in the hands of the physician, has become an indispensable auxiliary in the detection and diagnosis of disease." In many cases where long-known and well-attested methods of investigation have signally failed to detect the disease, the microscope has revealed important pathological truths.

For the purpose of familiarizing objects to the view of the learner, that would be of benefit to him in examining under the microscope, urinary deposits, we have prepared the CHART, which exhibits the microscopic appearance of a great number of such deposits. For the same purpose we would recommend a "full set" of already prepared specimens, which are to be obtained in almost any city, by viewing which under the field of the microscope a few times, there will be no need of ever mistaking the kind of deposit, any specimen of urine may present. By such practice you will also be much facilitated in the use of the instrument; and you will only have to let the urine repose a short time in a tall wine-glass, pour off the supernatant liquid, and place a drop of the under-stratum upon the glass-slide, when it is ready for inspection. Occasionally, however, it is necessary to resort to certain chemical re-agents, before the deposit can be satisfactorily ascertained.

In the absence of prepared specimens of "urinary deposits," THE CHART will be found a great auxiliary to the microscopic learner, in determining the kind or quality of matter he may have under investigation.

THE CHART should be spread to full view, in a place convenient to the instrument. Place the glass-slide containing the deposit upon the stage, adjust the focus carefully, survey the objects closely, then cast the eyes upon the CHART, and they will immediately catch the resembling *cut* or figure. A few alternate glances from the object to the Chart, and from the Chart to the object, will soon enable one to point out the complete resemblance. For it is a well-known fact, that even the most practical observer of anything, is not so apt to take note of objects or actions without his attention being directed to them, and hence the great value of "The Chart," in facilitating the progress of investigations.

EXPLANATION OF CHART;

AND

INDICATIONS OF THE DEPOSITS REPRESENTED.

FIG. I.—UREA.

A.—Represents the residue of healthy urine after the water, about 956 parts in 1,000, is all evaporated.

Healthy urine is of a light amber color, transparent, and of specific gravity about 1,020.

B.—Oxalate of Urea, which, when in excess in urine, indicates plethoric disease, with dyspeptic tendency.

C.—Nitrate of Urea, which, when in excess in urine, indicates derangement of the *cutaneous* function, with tendency toward both mental and bodily depression; and sometimes kidney affection.

A *deficiency* of urea, indicates *anemic*, dropsical, and hepatic diseases.

The average proportion of nitrate of urea, in healthy urine, is 14.5 parts in 1,000, but in some diseases it exceeds 30 parts in 1,000.

FIG. II.—CHLORIDE OF SODIUM.

A.—Represents crosslets and dagger-shaped crystals of the above salt, indicative of the presence of urea in excess; and found by evaporating the urine quickly.

B.—Chloride of Sodium, from water. It resembles oxalate of lime, but is distinguishable from it by being soluble in water, which the oxalate of lime is not.

C.—Chloride of Sodium, from urine evaporated slowly. It resembles oxalate of lime, also; but is soluble in water.

D.—Chloride of Sodium, resembling cystine, but is more soluble.

The average or standard proportion of chloride of sodium in healthy urine is $7\frac{3}{10}$ parts in 1,000.

CHLORIDE OF SODIUM

Fig 2

A.—Chloride of Sodium, evaporated quickly from Urine.
B.—Chloride of Sodium, evaporated from Water.
C.—Chloride of Sodium, evaporated slowly from Urine.
D.—Chloride of Sodium from Urine, resembles Cystine.

URIC ACID.

Fig 3

A.—Uric Acid, from Urine. Very Acid.
B.—Uric Acid. Rhomboidal form.
C.— do. Square form, inclining to Calculi.
D.— do. Rare form.
E.— do. Mixed with other Urates.

URIC ACID.

Fig 4

Varieties of Uric Acid found in the Urine of inflammatory [illegible] Rheumatism, &c., and in eruptive cutaneous diseases.

HIPPURIC ACID.

Fig 6

Hippuric Acid, ordinary form, abundant in urine of herbiverous animals.
Hippuric Acid, different form, abundant also in urine of some animals.
Hippuric Acid from Alcoholic solution.
do. do. from Aqueous solution.
do. do. from urine after action of hydrochloric acid.

PHOSPHATES.

Fig 7

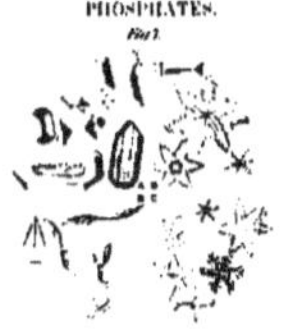

A.—Triple Phosphates, prism.
B.— do. do. penniform.
C.— do. do. stellar and foliaceous.
D.—Mixed Phosphates.

OXALATE OF LIME

Fig 8

A.—Oxalate of Lime, octohedral.
B.— do. do. when dry.
C.— do. do. dodecahedral.
D.—Oxalate Dumbbells.
E.— do. do. oval form.

ALBUMEN.

Fig 10

MUCUS.

Fig 11

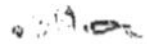

DIABETIC SUGAR

Fig 12

FIG. III AND IV.—URIC ACID.

A.—Represents uric acid crystals from very acid urine, and indicates the highest inflammatory state of the human system, inflammatory fever, so called, inflammatory rheumatism, gout, etc.

B.—Uric Acid, also, indicating a little less inflammation, with tendency to eruptive or cutaneous diseases, and found in measles, scarletina, etc.

C.—Uric Acid, square form. Indicates more local inflammatory action, with tendency to calculous affections.

D.—Same as C, with tendency to kidney disease.

E.—Same as C and D, with tendency to affections of the stomach, when mixed with oxalate of lime.

The average proportion of uric acid in healthy urine is only $\frac{4}{10}$ parts in 1,000; but in some diseases it exceeds two parts in 1,000.

FIG. V.—URATE OF AMMONIA.

A.—Represents the ordinary form of urate of ammonia, and indicates derangement of cutaneous functions, idiopathic fevers, (intermittent, remittent, etc.,) shingles, pemphigus, etc.

B.—Less common form; indicates same diseases as A, only less violent.

C.—Very rare form of urate of ammonia, indicates same diseases as above, with tendency to albuminaria, dropsy after scarletina, etc.

D.—Urate of Soda, usual form in urine. Indicates the same disease as above, with greater tendency to local inflammatory action, as in gout, scorbutus, etc.

FIG. VI.—HIPPURIC ACID.

This is not a normal ingredient of human urine, but is found therein after the administration of cinnamic or benzoic acid.

The urine of horses and cattle contain it in abundance.

A.—Ordinary form, from human urine, when the above acids have been administered.

B.—Different forms, found under similar circumstances to A.

C.—*Hippuric Acid,* from alcoholic solution.

D.—Same, from aqueous solution.

E.—Crystals of hippuric acid, after the action of hydrochloric acid on urine holding it in great excess.

This ingredient is not indicative of disease, but a knowledge of it may serve to prevent imposition.

FIG. VII.—PHOSPHATES.

A.—Represents *Triple Phosphates,* (prismatic) and indicate irritable state of nervous system, mal-assimilation of food, mental debility, etc.

B.—*Triple Phosphates,* (penniform) indicating the same as above, with tendency to disease of spinal marrow.

C.—Same, (stellar and foliacious) indicating disease of spinal marrow, and if excessive, hypochondria, etc.

D.—*Mixed Phosphates,* phosphate of lime, soda, magnesia, etc., indicates disease of brain and spinal marrow, with melancholy, hypochondria, etc., and if mixed, as in ammoniaco-magnesian phosphates, is indicative of some cutaneous diseases, such as shingles, herpes, zoster, etc.

FIG.—VIII.—OXALATE OF LIME.

A.—*Oxalate of Lime,* octohedral, indicating derangement of digestive organs.

B.—Same, when dry, indications same as above.

C.—Same, dodecahedral, indications same as above, with dyspepsia, inability of mental exertion, etc.

D.—*Oxalurate,* dumb-bells, same as above, sometimes disease of the bladder.

E.—*Oxalurate,* oval form.

These are essentially the indicative marks of dyspepsia, or derangement of digestive organs, and whenever found in the urine, in whatever form, all the varied symptoms of that many-headed monster may be confidently anticipated. (See Dyspepsia.)

FIG. IX—CYSTINE.

A.—*Cystine,* usual form, found in urine, and is indicative of scrofulous and cachectic disease, especially of the hereditary kind. (See Scrofula.)

B.—*Cystine,* crystalized from ammoniacal solution.

C.—*Vibriones,* animalculæ found in the urine of greatly debilitated persons of a cachectic habit, but more especially in that of those debilitated from syphilitic diseases.

D.—*Scaly Epithelium,* from the vagina, indicative of leucorrhœa, and irritation of vaginal mucus membrane.

FIG. X.—ALBUMEN.

Albumen is a coagulable lymph, and is one of the chief constituent principles of all the animal solids.

In albuminous urine is generally found large quantities of epithelium, the cuticle of mucus surfaces.

Dr. Gurnsey Smith says: "The peculiar appearance of the epithelial cells, will indicate the part of the genito-urinary mucus membrane from which the mucus was secreted."

A.—Represents *glandular epithelium* from the convoluted portion of the *tubuli uriniferi.*

B.—Pavement-like epithelium from the pelvis of the kidney. They are thin flat scales.

C.—Epithelium from ureters, being columnar or cylindrical.

D.—Epithelium columnar also, from the fundus of bladder.

E. C. F.—Epithelium from the trigone of the bladder, and from the bladder.

Moulds of the uriniferous tubes are often found in the urine, and Dr. Smith says: "They furnish valuable aids in arriving at a correct diagnosis as to the pathological changes which may be going on in the kidney." They are most commonly found in the urine at critical periods of acute inflammations, especially in scarletina, small-pox, pneumonia, etc. These, however, are fibrinous casts, (see Consumption and Scarletina,) some contain oily granulations, which are indicative of Bright's disease, (see Glandular Kidney.) *Albumen* is not an ingredient of healthy urine, but an important one of the blood; and whenever found in the urine, is indicative of *anemic* diseases, consumption, dropsy, Bright's disease, and such as pathologically exhibit *deficiency* of albumen in the blood.

FIG. XI.—MUCUS.

A small quantity of *Mucus* is generally found in urine, but it can scarcely be called an ingredient of healthy urine. When considerable, or in great excess, it indicates an irritation or inflammation of mucus membranes, especially of the bladder, urethra, vagina, etc., and contains epithelial cells.

A.—Represents mucus, found in urine, and indicates cystorrhœa, or inflammation of bladder.

B.—Represents mucus acted upon by acetic acid.

C.—Blood corpuscles, cohering.

D.—Blood corpuscles, separate. These are sometimes indicative of leucorrhœa if a female, but more generally of gonorrhœa when found in mucus urine.

E.—Represents oil globules, more indicative of diseased kidney.

F.—Represents epithelial cells, from the urethra, and is indicative of irritation of that part probably from gonorrhœa. Gonorrhœal mucus, abounds more in pus cells, than the leucorrhœal.

FIG. XII.—DIABETIC SUGAR.

A.—Represents *Torulæ Cervisæ*, which is only found in saccharine urine after it has undergone saccharine fermentation. Indicates diabetes, or disease in which sugar is found. (See Torulæ.)

B.—Represents a fungus growth in acid urine which contains albumen. Indicates weakness, lassitude, emaciation, and sometimes phthisis.

C.—*Spermatazoa.* Indicating seminal disease, nocturnal emissions, etc.

D.—*Peniculum Glaucum,* different form from B, the urine being acid and containing albumen also. Indicates a weakness of nervous system and general debility, Diabetis mellitis, etc.

FIG. XIII.—PUS.

A.—Represents *Pus Globules* as they appear in urine, and are indicative of suppurative inflammation, purulent absorption, or abscess of some part of the system.

B.—Pus globules acted upon by acetic acid.

C.—Large organic globules.

D.—Small organic globules.

E.—Coagulated albumen as found in Bright's disease, also in dropsy, etc.

The contents of abscesses of liver, spleen, and different other organs, have been known to escape into the kidneys and be discharged with the urine.

FIG. XIV.—LEUCORRHŒAL MATTER.

This figure represents the appearance of matters contained in the leucorrhœal discharges, and are a mixture of mucus, pus, blood, and epithelium, from the vagina.

Leucorrhœa is a disease of frequent occurrence, and the matter thereof may be detected in the urine of those laboring under that affection by their microscopical appearance, as here represented. *Vaginal* epithelium is always of the scaly variety. *Leucorrhœal* matter is to be determined from gonorrhœal, by the latter containing a greater number of pus cells. The increase of the cell elements and the diminution of the viscid matter, are indications of more virulent disease.

FIG. XV.—PUS, FAT, AND CANCER-CELLS.

A.—Represents *Purulent* matter from scrofulous disease, and when connected with *Cystine,* may be referred to scrofula of the hereditary kind.

B.—Represents *Fat* globules sometimes found in urine; they are usually fat cells filled with oil, and resemble those produced by mixing oil with water by the aid of mucilage. It appears to indicate a tendency to remarkable obesity; and is found only in the urine of persons of fatty degeneration of some or all of the parts.

C.—Represents *Uterine Cancer-Cells* found in the urine of females laboring under cancer of uterus or womb. "In cancer of the uterus," says Smith, "the microscopic examination of the discharges becomes highly important in arriving at an accurate diagnosis." "Cancer cells in such cases may often be detected in the discharges." The student should be careful, however, not to compound the columnar epithelium of the ureter, with the spindle-shaped cancer cells which they somewhat resemble.

FIG. XVI.—CANCER-CELLS.

A.—Represents the caudated cells which are of common occurrence in cancer tumors; and in cancer of the bladder are invariably present in the urine.

To know this, is of much importance, as there are noo ther means whereby the fact of such cancer of bladder, can be correctly established but by microscopic examination.

B.—Represents the concentric cell, as seen in cancer of the breast and ovaries.

The profession are nearly all now ready to acknowledge the necessity of a microscopic examination of those matters, to a correct diagnosis, and proper discrimination between pus, mucus and cancer-cells. "By the use of the microscope," says Carpenter, "the medical practitioner who is familiar with the aspect of urinary deposits, may determine on the spot, the nature of any sediment whose character he may wish to know."

GENERAL PRINCIPLES.

THERE must be first principles in medicine as well as in philosophy. Without a basis to stand upon; a standard from which to calculate; a starting-point as it were, the principles and practice of medicine, in fact the principles of any art or science can not be satisfactorily elucidated, intelligibly explained, or properly understood.

There must be a starting-point for every thing—a *zero* as it were from which to measure the *degrees* in excess or defect, or the kind of perversion that may exist.

Anatomy is the basis, the ground and pillar, the superstructure, the whole frame-work, upon which all rational medicine is built. It furnishes us with the "standard" of structural health; any irregularities or deviations from which, is structural disease.

Physiology gives us an account of the *action* of the organs and parts which anatomy only describes. The due and proportionate performance of these *actions* constitutes the condition and "standard" of functional health; any disturbance or irregularity in the performance of which, is functional disease.

A normal condition of a structural part is the standard from which any irregularity of structure is to be computed; the normal operation of any function, the standard of health of that function. Any excess or defect of structure or of function, is to be counted in *degree* or *kind* from these standards of normality.

Disease is defined by Williams to be "*a changed condition, or proportion, of function or structure, in one or more parts of the body.*"

Pathology is only a *description* of these changes—a description of the *irregularities* of anatomy and physiology, their causes, nature, and symptoms. "The investigation of these phenomena, and their reduction to general laws expressive of their conditions, is the object of pathology."

The *causes* of disease are divided into *intrinsic* and *extrinsic;* or those having their origin within, and without the body. Also into *predisposing* and *exciting* causes—of those which fall a little short of actual disease; and those which actually produce disease. The causes of disease are also divided into *cognizable* and *non-cognizable;* or those which we know by our senses, and those which we can only *infer* their existence by their morbific effects.

The difference between a *sign*, and a *symptom* of disease, lies in this—that a sign is that by which a disease is made known—a *symptom* being only the phenomenon which becomes obvious in the course of a disease. The one is perceptible to the *senses*—the other is dependent on physical properties.

The *vis medicatrix naturæ* is that power which unites fractured bones, and heals wounded parts—that resists injuries, as well as repairs them.

Predisposiny causes of disease are very numerous, comprising a long catalogue; and of all these, *debilitating* causes of predisposition to disease are the most numerous.

Previous disease—hereditary disposition, temperament, occupation, and the influence which *age* exerts in predisposing to disease—the several changes from early infancy to childhood, to puberty, to termination of growth, to *adult* and *old age.*

The predispositions of each of these, and their several pathological conditions, ought to be viewed. These being among the primary elements of disease, they contain many of the first principles of pathology.

Among the *exciting* causes of disease, are mechanical, chemical, non-alimentary metals, poisons, etc., togther with *excess* or *defect* of the alimentary.

The chief alimentary matters are the albuminous, gelatinous, oleaginous, and saccharine. *Albuminous* supplies the the albumen and fibrin of the blood; the *gelatinous* only supports it; *oleaginous* supplies the fat, textures, and secretions, and furnishes fuel for the maintenance of animal heat; *saccharine* supplies the material for the process of respiration.

These processes ought all to be examined in detail, and their *standard* qualities and pathological conditions studied; also the effects of heat and of cold upon the system, their pathological action, and the pathological conditions they produce—congestions, determinations of blood, and inflammations of organs, which are the results.

Endemic, *epidemic* and *infectious* causes of disease, are among the *non*-cognizable. Ague is endemic; cholera, epidemic; and small-pox infectious.

The causes, nature, and pathological conditions of these are interesting, indeed. They are all to be regarded as a blood poison, specific and different in effects.

Infection may be obtained by contact, through wounds, by exhalations from the body or breath, or by the air, through the lungs.

Infection is generally destroyed by heat, of one hundred and twenty degrees Fahrenheit, while cold also injures its vitality. Warmth, closeness and filth, increase its virulence.

Thus far we have only been viewing some of the *causes* of disease, or the elementary principles which enter into the formation of disease, or predispose to that result. This is necessary to the more ready arrival at the *nature* of disease. We will next inquire into the elements of disease itself.

Disease proper "is a changed condition from the natural structure or function of the body." There are *primary* elements—healthy and diseased—of structure and of function; and *secondary* elements of disease, composed of the primary in some combination.

The *varieties* of disease are comprehended in *degree* and *kind*. *Degree* includes *excess* and *defect*. *Kind* relates to *perversion*.

The property of contractility of muscular fibre may become *excessive* from over-feed of blood, and produce spasms, cramp, or tetanus. *Defective* contractility, by a want of due supply of blood to the muscular part, may result in paralysis, local or general, etc.

Tonicity differs from irritability in this, that its tendency is to slow and moderate contraction. "Its tendency is to keep muscles *in* their places when at rest, and *out* of their places when distended." *Irritability* and *tonicity* are the "standards" of functional health of muscles. *Excess* or *defect* of these, their pathological or diseased conditions. The same in the *nervous* structure. Insensibility, voluntary motion, reflex action and sympathy, are "standards" of healthy properties of this structure.

Excess, *defect*, or *perversion* of any of these properties, is disease—and hence, pathological. The pathological cause of *excessive* general sensibility, is an undue supply of blood to the parts of the cerebral mass concerned in sensation. That of *defective* general sensibility, is in the impeded circulation in the sensitive center of the nervous system. The pathoical cause of *perverted* sensibility must be sought for in the irregular supply, or *bad* quality of blood, which *supports* the nervous system. And so, in voluntary motion, reflex action and sympathy, all may be *excessive*, *defective* or *perverted*, according to *degree* or *kind* of pathological condition of the blood, and its circulation to or through the parts concerned in the disordered action. In all disorders of vital properties, these pathological conditions are plainly referrible to the changes in supply or quality of the blood to their textures, or parts concerned. The same is the case in diseases of *secretion;* and, like other functional properties, these may be *excessive*, *defective* or *perverted*, also. Each secretion has its own peculiar effects connected with its office; and these effects may be *forwards* on the parts to which the secretion goes, or backwards on the organ which secretes it.

But there is, besides *excess* and *defect* of secretion, a *perversion* of secretion. The normal action of secreting organs is the "standard," and these deviations constitute the diseases in *degree* and *kind.* *Perversion* is a deviation in *kind.* It is a morbid change in the secretion itself, as when the urine takes on qualities not belonging to it, etc. *Perverted* secretion may not only be caused by a deranged condition of the blood, or of its circulation, but it may have deleterious effects upon that fluid itself, and thereby become the source of different diseases, according to the parts engaged, and the kind of perversion.

And thus it is, that whatever property is exemplified, whether of irritability or tonicity of the muscular fibre—of sensibility, voluntary motion, reflex action, or sympathy of the nervous system, or of disease of secretions—the normal action of each is the "standard" from which to calculate the *excess* or *defect* in degree, or *perversion* in *kind.*

In whatever way these deviations from the normal standard may exist, and whatever effects they may produce, the blood will be found to play an important part in every pathological condition and action in the living body.

In fact, when we come to view the whole *make-up* of the human body, from the embryo to the full-grown man—and to consider that from the blood the form begins, the nourishment is given, and the growth is completed; as also from or by the same source injuries are repaired, and disease is removed; and that when the blood fails to carry on the process of life in any part, disease begins; and if not interrupted or overcome by the vital powers of that blood, death ensues; and that when death ensues the blood solidifies in part, and can not by any known process be reproduced out of the body, no more can the functions or properties of functions be carried on without this blood, we must conclude, with a writer of old, that "the blood is the life."

But let us examine still farther the secretions and excretions, and their effects upon the blood and upon the system.

Pathologists tell us that excessive secretions, if abounding in animal matter, may not only reduce the mass of blood, but also affect its composition.

If the urine be separated in unusual proportions from the blood, it must leave that blood modified.

Urine contains a preponderance of azote, and its excessive formation from the principals of the blood, would leave a predominance of hydrogen and carbon in this fluid.

But it might be said that the urine is not formed from the constant elements of the blood alone, but from the materials derived from food and from the decay and transformation of the tissues; whether, from the blood itself, or the co-operation of all these processes, a uniformity in the composition of this circulating fluid must be maintained, and if one of these processes is more active than the others, the blood must suffer by the excess of matters which the less active processes allow to accumulate in it. Williams sums up thus: "A flux of bile is either accompanied by a highly-loaded state of the urine, or by fever; in the latter case, the fever does not subside until the urine becomes very copious, or deposits an abundant sediment. The interpretation of this fact is: that the excessive secretion of bile disorders the composition of the blood. So long as the kidneys rectify this disorder, by separating in greater abundance the solid contents of the urine, no fever results; but if the kidneys fail in this task, fever ensues, and continues until they accomplish it; then a free secretion and copious deposit, are symptomatic of the decline of the fever."

But secretions may sometimes become *defective*, in consequence of a weakened state of the whole circulation, or of the secreting organ itself, or in a variety of ways.

"The distinctive materials of the urinary secretion appear to be positively noxious, and poisons the system if not separated from the blood. Thus—the sudden suppression of urine sometimes causes typhoid symptoms; and the amount of danger depends on the extent and suddenness of the diminution of the excretion."

In these cases, urea is found in the blood. The positively noxious property thus retained in the blood by suppressed urinary secretion, should always be taken into account when we are considering the conditions and treatment of idiopathic and symptomatic fevers, the constitutional effects of which are no doubt measurably due to this element.

"The changes in the blood, manifested in some such cases by its fluidity and by petechial appearances, may also be referred in part to defective elimination of effete matter thus retained, for it is when the secreting organs recover their power, and a copious discharge of highly-loaded urine occurs, that these appearances cease."

But there is *perversion* of secretion, and this often accompanies *excess* and *defect* of this process.

In febrile diseases the secretions from the kidneys especially, are prone to alterations exhibiting remarkable changes in quality, sometimes causing disorder even in their own organs, all of which will be referrible to changes in the blood itself, or in the supply of blood to their textures.

In examining the constituents of the blood and the changed condition in different diseases, it might be well to take a glance at the *properties* of the blood itself—for without a "standard" of properties as well as a "standard" of constituents, we can not measure these deviations, nor determine the extent of alterations.

This circulating fluid consists of red particles, called red globules, fibrin, or colorless globules, and *liquor sanguinis.* But for the better measurement of disease, as above stated, the "average natural proportions of these chief constituents of blood, in health," are given as a "standard," in a subsequent part of this work, any very considerable deviation from which proportion of constituents, is considered as disease.

Each of these parts having properties peculiar to themselves; although they may approximate their proportion, they may be defective in property.

"The red-blood disks seem to be the part of the blood on which its vivifying properties chiefly depend," and it is said that the vigor and beauty of the corporal frame is dependent more upon these red particles in the blood, than to any other constituent.

The obvious sign of this state of the blood is the florid color apparent in the lips, cheeks, gums, and other vascular parts of the body.

These red particles are defective in all anemic states of the blood, and all cachectic, scrofulous and tuberculous diseases, chlorosis, etc. The signs of this defect are, paleness of the parts naturally colored with blood.

Besides the *excess* and *defect* of these red particles, there may be *alterations* in these particles, as is evidenced in the readiness with which textures become stained in some diseases, as in petechia, etc., and also altered in form and size, and other properties.

Fibrin.—This appears to be the constituent which causes the coagulation of the blood, and the part which constitutes the buffy coat, and is considered to be the material by which textures are chiefly nourished and repaired. Its changes, therefore, must constitute an important element in disease. It exists in larger proportion and higher perfection, in arterial than in venous blood. It consists of congeries of fine fibres, with transparent bodies or globules scattered through them. In fluid blood we see the same pale corpuscles, but none of the fibres. The formation of these *fibres* is what distinguishes it from albumen. It has the peculiar property of solidifying in a granular mass, and this capacity is considered to be an attribute of inherent life—or closely connected with vital activity in the sanguiferous functions.

Its fibres, cells and granules, may be regarded as the rudiments of new living textures, and that it maintains the conditions of a most highly animalized material, which renders it fit for the peculiar properties of life.

The coagulation of fibrin is promoted by the contact of a rough solid; thus, by stirring fresh-drawn blood with a stick, the fibrin adheres in shreds to the stick.

The same property is exhibited within the body in the

deposition of lymph on the rough surfaces within the great vessels. The same cohesive property of fibrin causes it to aggregate in patches and films on the surface of membranes, and these form the basis of the construction or reparative process.

This coagulable lymph is carried to any and every part of the body, even into the minutest of structures, and there deposited freely when needed in any cases of lesion or solution of continuity of a part.

Deficiency of fibrin is of frequent occurrence in many diseases. In these cases there is a tendency to hemorrhages, to unmanageable oozing of blood from slight wounds or breaches of texture, and these are very difficult to heal, as also are fractures to unite, for want of the material for their nourishment and growth.

Albumen of the Blood.—It is this principle chiefly that gives the blood-liquor its thickness, and which renders it the more fit to pass along the vessels, preventing it from passing through their walls.

The deficiency of albumen in the blood is most remarkably met with in many diseases. This deficiency of albumen seems to be a chief constituent in dropsical diseases, and it is thought that all the cases of *anemia*, which are attended by dropsy, owe this concomitant, to a defect of albumen in the blood.

In cases of albuminaria, or disease of kidney, with coagulable urine, there is most remarkable deficiency of albumen in the blood, it being constantly carried off in the urine.

While we regard fibrin as the real vital or life-giving principle of the blood, and the nutritive and formative matter of the body, we know that there are those who incline to a different view, thinking, as Simon says, that they have cogent reasons, for placing it on the same scale as the extractive matters, and recurring among those elements which have arisen in the blood from their own decay, or have reverted to it from the waste of tissues. This view is taken of it by Dr.

Zimmerman, also. The cases which they produce, however, for proof of this position, appear to us, not to be well selected. In "the fact of *fibrin* being not only undiminished but usually increased in cases of repeated bleedings," may not this augmentation of vitality, or superfibrination of the blood, be a wise provision of nature to supply the waste in the body, occasioned by such loss of blood, and which remains unappropriated for want of this fluid to carry it thither?

And, in the fact of "the *fibrin* of the blood being proportionately increased during the progress of starvation," may not this state, also, be accounted for in the same way, or because of its non-assimilation, or non-appropriation to the parts requiring it?

And conversely, as regards the fact that "there is little or no fibrin in the blood of the fœtus," may not its excessive assimilation and appropriation necessary to that quick development, and rapid growth in the fœtus, keep the blood nearly *exhausted* of this life-giving principle?

In utero-gestation, there is relatively a more rapid formation or growth of the *living* body, by several hundred per cent., perhaps, than is ever accomplished in after life.

Now, as "these facts, derived from very different sources, appearing quite *in*explicable on the theory that fibrin is essential to the progressive development of tissues, and must be considered an excrementitious product, derived from the waste of tissue," we think there is nothing plainer than the opposite, and that these facts bear the proof in themselves, and are much more probable than the "*explosive* theory," as represented by some pathologists.

Fat.—The fatty matter of the blood is sometimes so very much increased beyond the normal standard as to give a milky appearance to the serum. The increase of fat in the textures is no doubt in some way connected with its excess in the blood. Exercise tends especially to reduce the fat of the body; while indolence, full living and good digestion, tend to increase it, and even to produce obesity. Unhealthy fat, however, commonly increases at the expense of strength, and is reduced in proportion as muscular power and activity is restored.

Saline Matter.—The saline matter dissolved in the blood, tends, according to experiments, to preserve the form of the red particles, and the fluidity of the fibrin. "There can be little doubt," says Williams, "that the thirst induced by salt food, is connected with an excess of saline matter in the blood, which causes a shrinking in size of the red corpuscles, and wherever they circulate, they attract by endosmose fluid from the textures and surfaces; thereby exciting that demand for liquid which the feeling of thirst is intended to supply."

Diminution of saline matters in the blood may cause it to coagulate in the vessels, and the red particles to become dissolved and altered. This appears to be the case in yellow fever, as reported by Dr. Stevens, and it seems that a defect of saline matter in the blood in cholera, has something to do in producing the lividity and collapse which characterizes that terrible disease.

Water in the Blood.—The proportion of water in the blood generally *increases* as that of the animal contents in the blood *decreases.* In all *anemic* diseases the blood is more watery than usual, as also after extensive hemorrhages or repeated venesections. The effect of this state of the blood is to cause a tendency to dropsical effusions, while a deficiency of water in the blood is exemplified in many diseases; in so extraordinary a degree in malignant cholera, that it can not circulate freely in the vessels, and hence this deficiency is the chief cause of the cessation of the pulse.

The Changes in the Blood. The first and most important change in the blood is produced by respiration—the process by which venous blood is made arterial and fitted for maintaining life.

According to Williams, this process comprises the absorption of oxygen, the removal of some carbonic acid and water, a slight increase of fibrin, etc. The absorbed oxygen, by its affinity for the hydrogen and carbon of the blood, perhaps, evolves heat, the renewal of fibrin supplies the expenditure of the plasma; while the removal of the carbonic acid is probably an excretion of noxious matter.

It is thought by some that this change in the blood from venous to arterial, is never carried on to *excess;* that the activity of respiration is always proportioned to the rapidity of the circulation, and the corresponding need of change in the blood. But in acute rheumatism there is great increase of fibrin in the blood, whether there be increased rapidity or not; hence the probability is that this *change* or conversion of venous into arterial blood, is sometimes carried on in *excess*. *Defect* of this change in the blood by respiration, is a very common element of disease, especially in affections of its respiratory apparatus; the phenomena of which is compounded by authors as follows: "1st, Accumulation of blood in the venous system; 2d, Diminution of blood in the arterial system; 3d, Deficiency of oxygen and excess of carbonic acid in the blood." There is, in these cases, not only a want of due supply of blood, but it is bad in quality, also. If, when the respiratory changes are *defective*, there is scarcity of fibrin in the blood—when we find the fibrinous element in *excess* in this blood, it seems highly probable that there is *excessive* respiratory changes in the blood, also.

Bile. The *defective* secretion of bile is often obvious in the yellow color of the serum and fibrin of the blood, and is connected with structural disease of the liver, or an imperfect action of that organ. It produces a general impoverished condition of the blood, with general cachectic state of the system.

Perspiration. "The perspiratory secretion contains lactic acid, and lactate of soda and ammonia, which it is believed proceeds, in part, at least, from the transformation or decay of the textures, particularly the muscles, which contain a preponderance of this acid. Hence, when this secretion is checked, these products may be retained in the blood and cause urinary disorders or cutaneous affections—thus, when the skin fails to secrete, an increased task is thrown upon the kidneys; if these organs fail in their task the lactic acid accumulates in the blood, and may produce either urinary or cutaneous affections according to the *pre*disposition of the patient. If there appears a copious deposit of this morbid matter in the urine the system may be relieved without much disturbance in the animal economy. The character and quality of the urine is a pretty sure criterion as to the affection produced, however; whether it terminates in inflammatory action of the urinary apparatus, or determination to the cutaneous surface, or, as is sometimes the case, an accumulation of the morbid matter deposited in the joints.

"In *Saccharine Diabetes*, the morbid matter is of a nature quite contrasted with that of gout or rheumatism, and the effect which each exerts is quite different. While the lithic acid in the latter disease has a tendency to accumulate in the body and to cause local irritations; the *sugar*, in the former, has no such accumulative tendency, but acts rather as a diuretic, passing rapidly away, carrying with it large quantities of water and the other constituents of ordinary urine." "The appearance of sugar in the urine," says Williams, "can not be considered otherwise than as a result of its presence in the blood, and is formed from starch and gum under the action of acids." How very readily may be detected these conditions of the system, by an analysis of the urine, where the patient has not been visited, these different diseases can be as readily distinguished by the different qualities of that fluid, or more readily than by all the other signs. In fact, it gives us the only positive declaration that can be made, and the only correct solution of these cases is made by an examination of the urine.

But the blood is changed in properties from the presence of foreign matters, such as poisons, excrementitious matters retained, etc. Some poisons act upon the body much quicker than others, and they all act much quicker upon an empty than a full stomach, as also poisons act much quicker in solution than when in a solid state. Any condition that favors quick absorption, favors the quick action of the poison upon the system. The most remarkably quick effects of poison are produced by prussic acid. It is said to produce its symptoms even in the act of smallowing it; and seldom are they delayed beyond one or two minutes."

How its effects can be produced upon the blood before the poison has reached the stomach, is explained by the fact, that the most ready contact with the blood in general circulation, is through the medium of the lungs; and that this substance is capable of producing, by *fumigation*, the most sudden effects. Its mode of introduction into the blood is most direct, and its effects as immediate. This poison has been detected in the *urine* in one minute after its introduction into the blood. *Pus* has been frequently detected in the blood by the aid of the microscope. The germs of many diseases are no doubt spread through the system through the medium of the blood by inhalation, by absorption, and by retained excretions.

The *first* object of the practitioner, in all such cases, should be to counteract the injurious operation of these matters; the *second*, to expel them from the system. How are we to know the antidotes to these, without knowing the kind of poison? By an examination of the urine we will be enabled to determine this point, or at least we will be much aided in our judgment in the case. But we do not always possess chemical antidotes, and especially those which can act on the foreign matter when in the blood, without injuring the blood itself; then our next object should be their expulsion from the system.

Here the excretory organs, especially the kidneys, must be called upon for the full performance of their functions. "Being the natnral emunctories through which foreign and offending matters are expelled from the blood, diuretic medicines will be found of utility here."

"Orfila found that the pernicious effects of small repeated doses of arsenic even, might be averted, by giving at the same time a diuretic medicine." Now we are fully satisfied that whatever poison be administered, or in what other way it may enter the system, and in whatever time it may take the same to act on the living body; whether in *three* seconds, as in some cases, or three days, or three weeks, as in others, the action of the poison is upon the blood, and the effects produced are through that medium.

Even where there is no perceptible interval of time between the contact of the poison and the production of its effects; and where there was scarcely an appreciable amount of the poison, the action is upon the blood, whether by shock, as an electric, thence upon the nerves, it matters not, the impression is made through this fluid. And when we see, not only these poisons themselves carried off in the urine, and find that the serious effects are thereby reduced, we can not but conclude, that the urine gives the most unerring evidence of the true nature of the case at the time.

Hyperemia.—The morbid conditions connected with *excess* and *defect* of blood in the vessels, have been somewhat noticed. The blood-making process is said to be ever on the increase in those whose digestive powers are in full activity, especially if these be indulged without exercise of the body. It may be in excess from too much blood being made, or from too little being expended. If, in either case, the great secreting organs fail in their proper function, disease is the immediate consequence.

Plethora may be *cthenic* or *asthenic*, or plethora with or without strength. *Sthenic* plethora comprehends a rich state of the blood and an active condition of the nutritive function, and has a tendency to cause general inflammatory excitement. In *asthenic* plethora, there is want of strength, of vigor and of action, notwithstanding the augmented quantity of blood in the vessels. It generally affects those weakened with age, excesses, or previous disease, and those in whom the excreting functions are imperfectly performed.

Anemia.—This term is applied to that condition of the system in which the predominant character is a *deficiency* of blood, together with a *deterioration* of its quality. Anything which withdraws or injures the blood, and especially the red particles, or anything which interferes with the formation of blood, tends to produce anemia.

This consideration of the system may be either general or partial; general, when the whole body is affected, and a general failure of the vital powers takes place; partial, when the defect of blood is more confined to a part of the system. Thus, whatever diminishes the supply of blood in a part, impairs the functions of that part, and *anemia* of the part is the result. Even a deficient supply of blood to the secreting organs themselves necessarily impairs their action, and softening and wasting of these organs is the consequence.

These conditions may be determined, in some measure, at least, by the character of the urine passed at the time.

Congestion.—Congestion may arise from an obstruction in the veins, which prevents a free escape of blood from them; or it may arise from *atony* or want of tone in the blood vessels themselves. It comprehends an *excess* of blood in a part, with motion of that blood *diminished.* If the veins be obstructed by a ligature, an excess of blood is retained in the part, because of the diminished motion of the blood, from the obstruction; and so anything which obstructs the blood in the veins leading from an organ, may produce congestion of that organ. Or anything which impairs the elasticity or tone of the blood-vessels themselves, or that may so enfeeble the action of the heart as to prevent the full circulation of blood through the whole circuit, by gravitation of the blood in the most depending parts, may cause congestions in parts that are lowest in position in the body. In proportion as the blood accumulates in one part of the body, it leaves the rest with less than its proper share, and hence weakness of these parts may ensue.

The stagnant blood of congestion not only becomes unfit for use in the animal economy, but it becomes a source of contamination to the rest of the blood, and causes cachexia of the blood at large when these contaminated matters are not promptly removed by the excretory organs.

Determination.—Determination of blood is different from congestion in this, that while there is *excess of blood in a part,* in both, in congestion the motion of that blood is *diminished,* in determination, the motion of blood is *increased.* In this condition there is generally an enlargement of the arteries as well as the veins, and the whole vascular plexus becomes the channel of a much-increased current both in activity and in volume. There is a stronger pulsation of the arteries leading *to* the part, and an increased quantity of blood in the part.

Determination of blood to any part, in a moderate degree, generally increases the natural secretions of the part. Thus, a determination to the kidneys may cause an increased flow of urine, and the urine itself may not only exhibit an increase of its water, but also of its acid matter, with more or less of the epithelial cells of the uriniferous tubes.

This result, occurring in the secreting organs or open surfaces, constitutes what is called *fluxes,* even "urinary flux." If it occurs in closed sacs or cellular textures, however, it constitutes dropsies. As flux and dropsy, then, may arise from a similar condition of the vascular system, they may be found to succeed to one another. The occurrence of dropsy is generally attended by a marked dininution of the urinary secretion, and a free flow of this often reduces dropsy. The occurrence of diarrhœa is also attended with a diminution of the urinary secretion, and a free flow of urine is a good indication of convalescence. In cholera, there is generally almost entire suspension of the urinary secretion, and among the first indications of convalescence is a free flow of urine.

When the vital powers of the kidneys have become reduced and unable to perform their vicarious action, in cases of excess of blood thrown upon internal organs; instead of separating the proper constituents of urine in sufficient quantities, the superfluous water accumulates in the blood, and by the quantity as well as quality, favors the effusion of serum. And when, in connection with this condition, there is loss of albumen from the blood, this loss still the more thins the blood, and thus facilitates the dropsical effusions. In these cases the albumen of the blood will be found passing off in the urine; the water of the blood, which should pass off as urine, is found in the cavities or tissues; and the salts which are wanting in the urine, are found in the dropsical fluid. This is the pathology of dropsy in a nut-shell.

Inflammation—In inflammation we have *excess of blood in a part*, the same as in congestion and determination, and like in determination, there is also a stronger pulsation of arteries leading *to* the inflamed part, but the motion of that blood is opposed by obstructions, and there is consequent stagnation of blood *in* the part. The difference between congestion, determination and inflammation, may be summed up thus: In *congestion* the capillaries are enlarged without any increase of the arteries. In *determination* the arteries and capillaries are both enlarged, and in due proportion.

In inflammation the arteries and capillaries are both enlarged, but the blood particles adhere to each other and to the walls of the vessels, and thus become more or less stagnant, until the obstruction is confirmed. Inflammation consists then in the increased motion of blood *to* the affected part with an obstructed flow *through* the part.

The *cause*, however, of the obstructed or retarded flow of blood through an inflamed part, has ever been a difficult question in the pathology of inflammation. "The coagulable lymph or colorless corpuscles of the blood, which is a constituent principle of healthy blood, and which has ever been considered as the *plasma* with which old textures are nourished and new ones are formed, these are found *cohering* in clusters and *adhering* to the walls of the inflamed vessels.

In congestion, these white corpuscles move slowly and sluggishly along the walls of the congested blood-vessels, and if not carried through by the momentum of blood, go creeping along the sides of these vessels slower and slower, until they finally stop one by one perhaps, so narrowing the path that the *red* particles of the blood can no longer pass, then these also stick and become so jammed by the current behind, that the whole of the vessels become of the deep red color of inflammation." These are some of the phenomena, and about the pathological condition of inflammation as presented by Williams.

Inflammation is the result of too long continued determination or congestion, and is always attended with more or less *effusion.* This effusion is generally at first a thin serum, and if the inflammation is slight it may remove it by unloading the engorged blood-vessels; this is called *resolultion.*

In some cases, after inflammation has continued for some time, there appears outside of the inflamed vessels white globules, very similar to those inside of the sevessels. These are called *exudation* corpuscles, and are supposed by some to be consolidated globules of fibrin. This *fibrin* is at first in a semi-fluid and ductile state; so that the motion or pressure of the inflamed surfaces draws it into bands or threads, or spreads it into films, it having an adhesive property common to glutinous material. The lymph thus effused is considered to be highly susceptible of organization, having the capacity of life itself. Possessing living properties, its materials arrange themselves into the bases of textures; and as the plasticity of this lymph depends much upon the good quality of the blood, to sustain the life of this texture, it must be supplied with healthy blood. For if the blood is poor in quality and the inflammation is of a low character, as it is apt to be when there is poorness of blood, these products of inflammation are less capable of organization, and are susceptible only of a low degree, resulting perhaps in tubercles. If the effusion of inflammation goes on to a great extent, pervading the adjoining textures, deranging their nutrition, and impairing their cohesion, softening of the textures take place. If these effused matters be retained, and new matters continue to be effused, the old texture thus becomes more compressed, and finally disintegrated and absorbed, and large pus-globules alone remain; this is *suppuration.* If the supply of blood be so stopped in an inflamed part that the part dies, the dead part may be separated from the living textures in the form of a *slough.* If the dead part pass into decomposition before it can be separated, this is called *gangrene.* If the inflammation be of a lower kind, the obstruction less complete, and the effusion more general, the nutrition of the textures only impaired, not arrested; it is called *induration.* These are called the *terminations* of inflammation. The *varieties* of inflammation are numerous indeed: Sthenic, asthenic, acute, sub-acute and chronic; congestive, phlegmonous, erysipelatous, diptheretic, hemorrhœgic, and scrofulous, together with the specific, as gouty, syphilitic, rheumatic, and gonorrhœal. These different varieties, each have their peculiarities whereby they are distinguished from one another, according to their prominent characters, and take on these names in accordance therewith.

Nutrition. The nourishment and growth of the body may be comprehended under the three heads of *increased, diminished,* and *perverted* nutrition. It is more or less a vital process, and the material on which the growth of a part depends, is the blood; and for the activity of the process the dependence is most upon the supply of arterial blood, the fibrin of the blood being the part which constitutes the basis of all textures. A rich blood favors nutrition, a poor blood impedes it, while a diseased quality of blood depresses it. These causes operate on the whole frame, but may affect only a part; in fact structural diseases are commonly partial from causes existing in the part. Whatever affects the nutrition of a part, affects its structure; whatever affects the circulation in a part affects its nutrition, therefore these diseases arise from differences in the quantity and quality of the blood and the variations in its character. *Increased* nutrition, or excessive natural growth of a part, is called *hyperthrophy*. When a part is necessarily more than usually exerted, it generally manifests the ability of accelerating its growth, and this circumstance is dependent, according to M. Paget, upon these three conditions: 1st, "The increased exercise of a part in its healthy functions; 2d, An increased afflux of healthy blood; 3d, An increased accumulation in the blood of the particular materials which any part appropriates in its nutrition."

Upon one or all of these three conditions depends the excessive natural growth, or hypertrophy of any part. In the case of *diminished* nutrition the conditions are somewhat reversed. There is a *wasting away* of the body or a part in all cases of *athrophy* or diminished nutrition. This may be general or partial, in a wasting away of the whole body, as in *marasmus*, or a dwindling away of a part of the body only, as a limb, or an organ.

The cause of *Atrophy* is divided into the circumstances which promote decay, and those which impair or prevent the reparatory nutrition.

A part may become atrophied by disease or want of sufficient exercise of its functions, for want of sufficient or defective supply of blood; or, because of defect in the formation of fibrin and albumen of the blood, the plastic materials of nutrition.

Whether hypertrophy or atrophy, the urinary secretion always plays its part in the removal of the different kinds of debris, according to the nature of the case, or exhibiting in its qualities the condition of the blood at the time, and consequently the nature and condition of the part.

In *hypertrophy*, like in hyperemia, in general, there is usually a diminution of this secretion and of its most characteristic constituents. Being secreted in less quantity than usual, its specific gravity will be higher, and its color darker than natural, and will sometimes approximate in composition the urine of inflammation.

In *atrophy*, however, the urine is different. In these cases it will often be found to contain products resulting from the decay of the textures. In marasmus, for instance, "from the excessive secretions or drains from the body, there is always proof of accelerated decay of textures, manifested in the increased amount of urea excreted by the kidneys."

Besides the structural lesions of hypertrophy and atrophy, there is *indurution*, *softening*, and the different *degenerations*, all of which are called *alterations* of the textures.

In addition to these, there are *Deposits* in or upon textures, which are new matters added to these *alterations*. By *deposits* are meant the matters which result from an *overflow* of the nutritive material beyond what is necessary to nourish the textures themselves. The basis of these is the fibrinous matter of the blood, and are divided into: 1st, Those having a high organizability; 2d, Those having a defective organizability, and 3d, Those destitute of organizability.

There is every gradation in these from the highest capacity of inherent life belonging to organized bodies, to the most degenerated state of the plasma of the blood found in the non-fibrinous tubercle itself.

And we will be much aided in our judgment of these grades, by a knowledge of the condition of the blood, and the different substances imparted to the urine in each of these different conditions.

Signs.—The difference between the signs of health and the signs of disease, is determined by our knowledge of what is usualin health, and this knowledge is to be derived from our experience of a healthy standard.

There are many signs and symptoms of disease, and it behooves the medical practitioner to avail himself of *every* possible means to ascertain the true nature of all cases he may undertake to treat. And, while we are claiming for the urinary secretion more attention than is generally given to it, believing it to be of more value in making up a diagnosis in every case, than is usually acccorded to it, we do not decry, or wish to detract the value of any other mode. The state of the pulse is considered to be an important source of symptoms. A slow pulse indicates a deficient excitement of the heart, while a frequent pulse indicates the reverse; a strong or weak pulse implies an increased or diminished strength of the heart's contractions; a hard or a soft pulse, an increased or diminished tonicity of the arteries; and so on with all the different states of the pulse; they all indicate, denote or imply, certain states or conditions of the circulation; but of that uncertain and variable character which will not alone render a criterion in disease. The same may be said of the appearances of the tongue, as also of the alvine evacuations, and all secretions and excretions.

In a person whose growth is stationary the quantity of food passing into the body in a given time, is exactly equivalent to the quantity of matter passing away in various *excretions.* "The daily six pounds, more or less," says Simon, "of meat, bread, coffee, salad, water, etc., exactly correspond to the quantity of fæces, urine, sweat, expired carbon, etc." Excepting a very small portion which may pass through unaltered, the whole of the food is dissolved in the body, and the greater portion is absorbed into the circulation, to renew the tissues. The blood as it circulates gives to each organ the means of repairing itself, by furnishing it with material of new growth. But the blood itself must *live* and *grow*, and its life and growth must precede all other life and growth in the body; for first *it* grows, next other organs grow at its expense.

The blood possesses the power of washing away, as it were, from each organ, in a dissolved state, whatevar elements of its tissues have become wornout and useless. These are passed off in the excretions. If you take into the stomach a little rhubarb or iodide of potassium, and find traces of the drug in a few minutes in the urine, you know that it must have traversed the circulation, been absorbed, dissolved in the serum, and literally to have formed part of the blood during the time elapsed between its absorption and its discharge. The same thing occurs in the natural process of organic conversion. In the chemical changes which occur in the active tissues of the body, the disintegrated particles, in order to reach the kidneys, for elimination, must first form a part of the stream of circulating blood. This passage of particles through the circulation, and its elimination, by the kidneys, is extremely rapid, in some cases, especially in that of some poisons, the elimination by that organ being as rapid as its absorption by another—a wise provision of nature, indeed, for the preservation of the integrity of the blood, and the safety of the body.

BLOOD.—" The blood is the life." The heart is the fountain-head from which this vital fluid issues. The right auricle of the heart, is the part first seen to pulsate in the embryo, and in death it is the last to retain its motion.

The commencement of the circulation of this vital fluid, then, may be regarded as the beginning of active life, and its cessation, the end; the blood itself being the support of all the vital properties, or phenomena belonging to animal existence.

In the constituent elements of healthy blood, are contained all the vital properties of the elementary solids of the body, while in the morbid elements of the blood may be traced all the elements of disease.

All diseases, therefore, are manifestly first in the blood. And, as from the blood is secreted the urine, through the medium of the kidneys, an acquaintance with the constituent principles of the blood itself, both in health and disease, is of vital importance to the student of medicine, and indispensable to the proper understanding of the Uroscopian System. It will not be our province, however, to give a minute description of these, nor to describe the process of analyzation of the blood, as many valuable works have already been written thereon, but only to give a mere statement of some of the more important facts that seem to be called for in the elucidation of our subject. And the first of these will be found in the adoption of the average natural proportions of the chief constituents of the blood in health.

The many and repeated analyses of the blood, by the celebrated Lecanu, which enabled him to form a "standard of proportions" of the chief constituents of healthy blood, the correctness of which, all the analyses of both the celebrated authors, Andral and Gavarret, have but the more fully substantiated, and which standard they adopted, will be a sufficient guarantee for its adoption here.

This proportionate standard is as follows:

Red globules, ...	127	parts	in 1,000.
Fibrin, ...	3	"	"
Animal matter in the serum, .	72	"	"
Salts, ...	8	"	"
Water, ...	790	"	"

Now, the above being the average natural proportion of the chief constituents of the blood in health, any considerable deviation from this standard may be regarded as disease; and the extent of such deviation a tolerably-fair measure of the extent of the disease.

The red globules appear to be the part of the blood upon which its vivifying properties chiefly depend. Their proportion is from ten to twenty parts in one thousand, less in females than in males, an important fact necessary to be borne in mind during examinations of the urine.

These red globules are always in excess in sanguinous plethora, in some instances rising, it is said, to one hundred and eighty-five parts in one thousand of blood; while in other instances, such as after great loss of blood or in chlorotic and anemic diseases, they have been found by Andral, to be reduced even to twenty-seven parts in one thousand of blood. The former represents force and strength, the latter feebleness and debility.

The remarkable diminution of the red particles of the blood, which sometimes takes place, is generally due to the draining away of the albumen of the blood, through the medium of the kidneys, which albumen is always perceptible in the urinary secretion, and may be detected by the appropriate chemical tests, or microscopic appearance. (See Chart, Fig. 10.)

When the red particles are in excess in the blood, there will be no albuminous substance to be detected in the urine, but generally an abnormal quantity of fibrinous matter or other properties exhibited in that secretion, indicative of an inflammatory condition or plethora, in extent according to such abnormality in quantity or quality of the substances thus cast off in that fluid.

In Williams' Principles of Medicine, page 158, he says: "The blood is probably the chief seat of the morbid poisons which excite various contagious, epidemic and endemic diseases. Probably, too, it is the hot-bed in which some of them are propagated, whether by seeds, *ova*, cell-germs or parasites; and it is through changes in its composition, (the blood) that many of the destructive effects of these poisons are produced."

In the treatment of foreign morbific matters in the blood, the same author says: "The two indications which present themselves are—first, to counteract the injurious operation of these matters; and, second, to expel them from the system. The first of these indications is followed when we give stimulants to overcome the depressing influences, etc.; the other indication is more generally pursued, although little recognized by practitioners, to-wit: to expel the offending matter from the blood. The excretory organs, especially, the kidneys, and alimentary canal, are the principal natural emunctories through which foreign and offending matters are expelled from the blood, and hence the great utility in aperients and diuretics, in the treatment of fevers and other diseases connected with injurious matters in the blood."

Orfila found that the pernicious effects of small and oft-repeated doses of arsenic, might be averted, by giving at the same time, a diuretic medicine. How often do we see fevers and other serious ailments, carried off by even a spontaneous diuresis. And this is effected by the removal of the deleterious matters from the blood, through the medium of the kidneys and urinary organs.

The great variety of changes that may occur in the blood, is not more than equalled in the variety of diseases that afflict the human family. Even the coloring matter of the blood is evidently altered in many ways, and is changed in many diseases. In the worst forms of scurvy, it is said, by Mead, to be changed to a dark-brown, or green color; and in the Walcheren fever it was described as being pitchy black.

In the worst form of cachexia, the blood is not only poor but perverted, exhibiting various shades of purple, brown, and even greenish colors. All these have their significance, pathologically, and must be taken into account. Also, the positively noxious properties which excrementitious matters retained in the blood, are known to possess.

One of the most interesting facts connected with the pathology of the epidemic fever which prevailed in Edinburg, in the year 1843, is said to be the discovery of urea in the blood. "The existence of urea in the blood," says M. W. Taylor, "in other cases has been inferred from occurrence of disorders of the nervous centers, which we know to be the consequence of its undue accumulation in the circulation. These phenomena have been observed in those cases in which, from some cause or other, *the daily discharge of urine has undergone material diminution.*"

It is well known that the blood in dropsical affections exhibits a deficiency of the usual proportion of albuminous matters, and that in such cases there is found in the urine considerable quantities of that substance. So general has a knowledge of this become, that nearly every physician now tests the urine in dropsy, and treats the case accordingly.

Dr. Blackall, in speaking of this circumstance, says: "Let me add once for all, that in the instances of cure (of dropsy) by digitalis, the improvement of the urinary discharges is simultaneous with the relief of the other symptoms, certainly not subsequent to them, but always among the earliest good signs."

Having given the standard of the chief constituents of the blood in health in page 59, an excess of any of these constituents over the proportions there given, may be considered abnormal in quantity, and therefore declarative of one form or kind of disease, while a deficiency of the proportionate quantity of the same constituents would be declarative of a reverse kind. Thus, "in some forms of disease, as for example in anemia and chlorosis, the proportion of *water* is usually much greater, and has been known to amount to upwards of nine hundred parts in one thousand; while in certain other pathological conditions, on the contrary, the blood is found to contain considerably less water than is present in the healthy blood; as in cholera, for instance, where the blood is so rich in solid matter as almost to resemble jelly in appearance, it has been known to contain not more than four hundred and eighty parts of water in one thousand of blood." (Bowman.)

The same author says: "In disease, especially some forms of fever, the proportion of red corpuscles sometimes increases considerably, and has been known to amount to one hundred and eighty-five parts in one thousand of blood (one hundred and twenty-seven being the average), while in certain other affections long known as having been attended with great poorness of blood, the proportion of red corpuscles frequently does not amount to more than sixty or seventy parts, and has been known to be as low as twenty-one parts in one thousand of blood."

The average proportion of *animal matter* or albumen in healthy blood being seventy-two parts in one thousand, according to the given "standard," in some diseases it is as high as one hundred and thirty-one, while in others, as in Bright's disease, it is sometimes as low as fifty-five parts in one thousand. The amount of deficit of this substance in the blood in this latter disease, is always connected with a corresponding amount of the same substance, discoverable in the urine. And it is by the quantity of this substance being thus cast off that the magnitude of the disease may be quite correctly estimated."

Minute traces of *urea* are thought by some probably to be always present in healthy blood, but the quantity is very small, while in some forms of disease, especially those in which the functions of the urinary organs are to any extent interfered with, the amount of urea is found to be considerably increased, "and may frequently be met with," says Bowman, "in a sufficiently large quantity to be weighed."

The average proportion of *fibrin* in healthy human blood being three parts in one thousand, it has been known to vary from a mere trace to upwards of ten parts in one thousand of blood. A considerable increase in the amount of fibrin in the the blood is always found in every form of imflamatory disease, and a corresponding deficit of that constituent in reverse diseases, or those of a non-imflamatory or anemic character.

The average proportions of salt being eight parts in one thousand of healthy blood, in scurvy and some other pathological conditions, their amount has been found increased to as much as eleven or twelve parts in one thousand, while in other diseases the amount many times falls much below the healthy standard.

Learning the chief constituents of healthy blood, and the relative proportion each bears to the other, and that each is subject to changes of proportion, either in excess or deficiency, and that such changes are characteristic of different diseases, the necessity of searching for the source of removal of said constituent in the case of its deficiency, or the cause of its redundancy when in excess, will be apparent to every inquiring mind.

When in connection with these we find that in anemic or chlorotic diseases, in which the blood is poor in corpuscles and excessive in water, the urine is scanty and loaded with coloring matter, indicative of a retention in the system of the one, and a drainage from the system of the other; and when in dropsy or albuminaria, the blood is found to be deficient in albuminous matter, while the urine is loaded with that substance; as also in inflammatory diseases, in which the blood loses its fibrous property, and said substance is found to be deposited in the urine; and, as also, the same general rule is observed in scorbutic and other diseases, in which the salts have been found in excess in the blood, and a corresponding deficit of the same in the urine, and vice versa; when we find all these facts and numberless others of a like significant character, we feel confident in the correctness of the general principles, and hope to make them valuable in practical results.

But apart from these facts, which are the result of the experience of many eminent authors upon the subject, another circumstance connected with the relative conditions of the blood and urine deserves our attention, namely: morbid blood, containing abnormal ingredients.

Blood not only becomes morbid by an excess or deficit of any one of its natural constituents, but it is subject to morbidity by its reception of ingredients foreign to its composition. And in this condition of the blood, a different state of circumstances exist, and a different rule of action is set up. The ingredient contained in the blood being abnormal, or being foreign to the composition of the blood, the kidneys being the emunctories whereby foreign or morbific matters are removed, portions of these abnormal ingredients will be found in the discharges therefrom. And *commensurate with the quantity of said abnormal ingredients in the blood*, so will the relative quantity of the same substance be discovered in the urine.

Thus, "the blood of patients suffering from Diabetes, appears mostly to contain a very sensible amount of sugar, and this substance may in such cases very readily be detected in the urine." (See Diabetes.) And so it is in jaundice and some other affections in which the functions of the liver are interfered with, an accumulation of biliary matter is found to have taken place in the blood, giving to the serum a more or less decided saffron or orange-brown color, "which is due to the peculiar coloring matter of the bile," says Bowman, and which may be readily detected in the urine by appropriate chemical tests. And thus it is with pus, fat, or any poisouous substance, of whatever kind, that may have entered the blood, whether by food, drink, medicine, malaria, or the decayed particles of other parts of the human system, and which it is the office of the kidneys to secrete and the urine to carry off.

The human system is continually undergoing the process of reparation and decay. The elements of food being absorbed by the lacteals, reach the right side of the heart, and being exposed to the influence of the air in the lungs, become converted into blood. "From the blood all the tissues of the body are formed, and the waste of the animal structure supplied." Before this nutrient substance can be deposited, room must be made for it by the removal of old and exhausted material. The blood is sent out from the heart through the arteries, highly charged with the necessary elements for the full supply of all the wear, tear, and decay of this body, as is evidenced by its bright and lively florid colors, ; and on its return through the veins, it takes up the decayed particles from every tissue, thus virtually making room for its load. That the blood thus regularly unloads itself of its *vivifying* properties on its outward-bound trip from the heart, and gradually re-loads on its return through the veins with decayed particles, poisons, or whatever *debrs* or deleterious substances it may have met on the way, is fully evidenced from the great change of color and other properties it undergoes in the round, being darker, almost amounting to a black as it approaches again the heart.

A paucity or excess of one of the constituents of the blood, or its contamination by morbid influences or foreign substances, constitute the essential cause of every disorder which afflicts the human family. And the peculiar symptoms which enable us to distinguish one particular disease from another, is the result only of such paucity or excess, or morbid influence on the blood. And the urine being secreted from the blood only, under all these circumstances, in consideration of the office which it is detailed especially to perform, to-wit: "removing from the blood superfluous and unhealthy matter," it is plain that these differences must be characterized in this eliminating fluid.

Now, as the real cause of a departure from health is first manifest in the blood; and as such cause, whether by change in the constituent principles of, or foreign substances in the blood, is early made known in the urine, even before the symptoms by which we generally characterize the disease, is made known upon any other part of the system; or rather, as Blackall says, "as the change in the urine is among the earliest symptoms of certain diseases," and as "the state of the urine is also generally among the *first* and most *convincing* signs of *improvement* in disease," and as "without any particular reference to the other symptoms, and even when the patient has not been visited," (says the same author,) "the state of the urine itself furnishes us an important indication of the disease," and as the kidneys excrete the only substance containing particles from *all* parts of the system in sufficient quantities for the practical purpose of diagnosing diseases, and as the condition of those particles exhibits the *true* condition of whatever part they proceed from, will any one say that it is not of the utmost importance, in all cases, to examine the urine, in order to arrive at the true cause of disturbance in the animal economy? For there is no change that can take place in the blood, either in the constituents of the blood itself, or by foreign matters, which have merely entered its circulation, but that evidence of the same, or traces of the substance is to be found in the urine, even poisons, arsenic, etc., which, according to Orfila, "are carried off largely by the renal secretions."

The dark color and thick consistence in blood in Asiatic cholera; the paleness of the blood in chlorosis; the whitish leather-like pellicle which covers the blood in pleura-pneumonia; the yellowish appearance of the blood in jaundice; the dissolved state of the blood in scorbutus; the reddish, dirty-water appearance of the blood in anemia; the deficit of albumen in the blood in dropsical and other diseases, etc., etc., are so many evidences of the different morbid conditions of the blood in different diseases, that we do not hesitate to pronounce that all diseases manifest themselves first in the blood, and that the continuance of disease is cotemperaneous with the continuance of the altered condition of the blood; and that the cure of any disease will only be effected by a restoration of the blood to its normal condition.

The whole amount of blood in the human body has been reckoned to average about thirty pounds; the quantity projected by the heart at each systole, to be from one to two ounces. Therefore, calculating the pulsations to be seventy-five per minute, it would take only from three to seven minutes for all this blood to pass through the heart. Now a constant metamorphosis is going on in the living blood. When it ceases to undergo this metamorphosis, it dies; but the blood is not the only portion of the body that undergoes this change; every organ and tissue is subjected to a similar metamorphosis, which is presented to us under the general phenomena of nutrition and consumption, or reparation and decay, and which is dependent on, and effected by the blood alone.

But, since the various tissues present a different chemical composition, and since the different organs separate different matters from the blood, it is obvious that they can not all modify this circulating fluid in the same manner.

In the conveyance of nutriment to the various parts of the organism, the blood, in its passage through the capillary net-work, permeates all organs and tissues, and their cells take up from the plasma those substances which they require for nutrition, and restore to it those which have become effete, and are no longer adapted for the process of nutrition.

Now, in view of observations already made, in regard to the position of the kidneys in the human system, the great caliber and short course of the renal artery, the connection with the liver, spleen, omentum, etc., and that, according to some, the fifth part of the whole blood passes through the kidneys, that at each systole of the heart five or six scruples of blood are driven into them, together with the peculiar structure and chemical constitution of these organs themselves, permeated as they are by such an extremely abundant and dense capillary net-work, and such very delicate venous twigs, so closely encircling their excretory ducts that the tissue is brought in contact with this blood at every point and in every direction, we must infer that the blood thus may not only undergo a much more rapid metamorphosis in the kidneys themselves, than in any other organ of the body, but that the kidneys the more readily separate the metamorphosed particles from other organs and tissues with which this blood is surcharged than is likely to be done by any of the other excretory glands of the system, and that by an examination of this secretion the mind of man is enabled to penetrate more deeply into the pathological conditions of the human system, and to learn the material deviations from a normal state with greater accuracy than can be attained in any other way.

URINE.

"The urine is by far the most compound of all the secretions," says Dr. Aikin, "draining off, as it does, particles of every tissue which the absorbents take up from all parts of the body, as wornout, redundant and useless."

The position of the kidneys in the human system, the connection of the right kidney to the liver, which rests on it, as it were, and the left on the spleen, and both, to the supra-renal glands, by a cellular tissue, gives to these organs a superiority of position over many others, in their anatomical connection as well as physiological consideration.

The blood-vessels of the kidneys are very large in proportion to the size of these organs themselves, the renal artery being one-eighth the size or caliber of the great aorta of the body. It is also very short compared with any other artery of the same size. And from this, short course and the peculiar distribution of its branches, together with the great size in proportion to the kidneys, it is peculiarly well adapted to that quick and copious secretion of fluid which the kidneys so readily separate from the blood.

The renal artery and its branches so freely communicate with the veins, and so readily open into the excretory ducts or conical nipples, from which the urine being now separated from the blood, trickles into the funnels, thence into the basin, thence through the ureters into the bladder, and by this short course, afford so ready a passage for the evacuation of so large a quantity of a fluid which has been secreted from so large a volume of blood, the kidneys can not but be regarded as the great balance-wheels, as it were, the regulators and the purificators of the circulatory system, but their secretion, as the *indicator* of the health and strength of the whole animal economy also.

This character of these organs is more fully strengthened the more we examine their adaptation to these purposes, and the further we pursue the investigation into their physiological condition, and connection with the circulation of the blood; whether in view of the offices they fulfil ordinarily, or their ability to act as a compensating functionary under more than ordinary circumstances.

"The kidneys," says Dr. Aikin, "constituting as they do, important emunctories or channels by which superfluous or hurtful matters are discharged from the blood, it is necessary that their function should be carried on without interruption, as is found to be the case in health—for the constant deposition of new, and the absorption of the old materials, demand the unceasing evacuation of the old wornout particles from the general mass of the circulating fluids." Neither can the function of the kidneys be suspended for any length of time without injury to the constitution; for when this secretion is suspended, or its evacuation obstructed, great distress is sometimes experienced in the whole system; because of all the fluids the urine is that which hastens most speedily into putridity, symptoms of decomposition taking place, in some instances, even before the extinction of life.

And, when death does take place, from a suspension or obstruction of this secretion, "the body rapidly passes into a state of offensive putrefaction, indicating how important to the health and well-being of the whole frame the due performance of the function of the kidneys necessarily is."

"The urine is a fluid," says Markwick, "composed of certain *effete* animal and saline matters which have been separated from the blood by the kidneys," and "as it carries off from the system," (says Aikin) "the *debris* of the body, so does it vary in different states of the constitution."

When any one organ becomes diseased, its vitality being lessened, it will have less power to resist the tendency to decomposition, hence there will be more of these diseased particles passed from this organ than from the healthy parts. And, as the different parts of the system are of different composition, colors, etc., so much so that almost any person can distinguish, merely by their appearance, the lungs from that of the heart, liver, brain, kidneys, etc., or one of them from another, so the *refused* fluid, or *diseased particles* of these parts are also different one from another; and are as capable of being distinguished from each other, by their ocular appearances, chemical tests, or microscopic characters, and the different qualities of each may be as clearly ascribed to its different locality or organ, by a practical Uroscopian, as can a practical chemist, by an analysis of the waters of the earth, determine the presence, in the vicinity of said water, of salt, lime, sulphur, iron, etc., or a microscopist, by the *lacunæ* and *canaliculi*, their shape and size, determine to what class of bird, beast, reptile or fish, any given piece of bone belongs.

The waters of the different medicinal springs, the Spa, Seltzer, Harrowgate, Bedford, Cheltenham, Saratoga, etc., are readily distinguished by their appearance, color, taste or odor, and the chemist is enabled to determine the different ingredients composing each, and the definite proportions of the same; which ingredient predominates, and to what extent is the deviation from pure water. The same may be said of a practical Uroscopian.

Dr. Simon says; "The analysis of urine seldom presents any very great difficulty. Many of its constituents may be detected with ease, unless, as is sometimes the case, they exist in very minute quantity." It is something like that of the mineral waters, some of the constituents may be at once recognized, even without any test, others by the addition of a simple test, while the presence of others may require to be separated and isolated, and viewed under the microscope. "During disease the urine may undergo numerous modifications, both in its physical character and its chemical constitution." The chemical changes may be reduced to one of the following forms:

1st. One or more of the normal constituents of the urine existing in larger quantity than in healthy urine.

2d. One or more of the normal constituents existing in less quantity than in healthy urine.

3d. A normal constituent absent.

4th. The presence of substances that do not exist in normal urine.

Of the first, the deep-brown color will indicate the abundance of solid matters. Of the second, the pale urine, or more water-like, will indicate their deficiency; but the common urinometer is sufficiently accurate to determine the excess or defect in either, by giving the specific gravity.

The practitioner should not only be able, by examination of the urinary secretion, optically, chemically, and microscopically, if need be, to determine the character of a given specimen, whether the constituents of healthy urine be present; if so, whether in due proportion; if not, in what the deviation consists, but also, the extent of such deviation from the natural standard—which would at once give a clue to the nature of the disease, or the cause of departure from health.

It has been stated by Dr. Buchan, in his "Domestic Medicine," "that the passions, the atmosphere, the food, clothing, state of the evacuations, and numberless other influences, tend to so change the quality of the urine as to render it an uncertain criterion in disease," etc. Now, these same influences tend to, and *do* change the condition of the patient at the same time, and precisely in the same ratio of the change in the urine. Is it pretended that a violent fit of anger will change the appearance of the urine, without producing any change upon the nervous system? Or, that a sudden change in the atmosphere will produce a marked change in the appearance, quality, or constituent principles of the urine, without producing any impression upon the constitution or some one or more of her organs? And the same questions may be asked in reference to food, clothing, state of evacuations, etc., etc. All these produce certain impressions upon the constitution, by making certain changes in the circulating fluid, the blood; hence the change in the character of the urine, which change in the urine is the index to the character of the disease, or peculiar condition of the system at the time.

The statement of Dr. Buchan was copied by Dr. W. Beach, and published in his "Family Physician," a number of years ago. Afterward his attention was called to the subject by the author of this work, in a letter written to him in the year 1844. In his answer to said letter, Professor Beach says: "With regard to what I have said in my work, page 59, etc., I am now willing to stand corrected on the point referred to in your letter. I may have been too sweeping and wanting in discrimination. If so, I am ready to retract."

Several years after the above was written, and when he had more fully investigated the subject, he publishes the following retraction, in his more valuable work, "The American Practice," Vol. I., page 175: "To the class of physicians who are called Uroscopians, and depend mostly on this secretion (the urine) to form their diagnosis, we owe an apology for ridiculing their pretensions. We have since learned, though no doubt some are extravagant, that they *are* enabled to discriminate between diseases by an examination of the urine; and the greatest physiologists in Europe now pay particular attention to the quantity and quality of the urine in disease."

Disease is the antagonist of life, and the life of a person is the blood—the complete destruction of the blood being death. Every attack of disease is an attack upon the blood, and upon the repulsion of such attack depends the life of the body. Therefore, to arrive at the precise condition of the blood in any disease, should be the first object of the physician, as by such knowledge he is put in possession of the *data* whereby its repulsion is best accomplished. That the urine is secreted from the blood, and that the blood and the urine are both definitely and almost simultaneously altered in composition, in nearly every different disease, that any deviation from the natural standard of the chief constituents of the blood in health involves a corresponding deviation from the natural standard of the chief constituents of the urine in health also, are facts which seem to be well established by theory as well as practical experiments and observations. To arrive, then, at a correct knowledge of the different proportionate ingredients of urine, in its various deviations from the healthy standard, will put us in possession of a tolerably correct knowledge of the state of the blood at the time.

The average proportion of the constituent principles of urine in health, is given in the following table, as approximating a standard at least, which is sufficiently applicable to all general purposes, and sufficiently specific for all practical purposes; the average density, or specific gravity, being placed at 1,020 (water being 1,000), in adult males, a little less in females, and still less in children, according to age, etc.

TABLE.

Water, - -	956		parts in 1,000 of Urine.
Organic Matters,	30	Solid Matters.	" " "
Fixed Salts, -	14		" " "

The Organic Matters are composed of the following ingredients, and in about the following proportions, to wit:

Urea, - - -	14.5		parts in 1,000 of Urine.
Uric Acid, -	.4	Organic Parts.	" " "
Extractives, -	14		" " "
Mucus, - -	.2		" " "
Muriate of Ammonia,	.9		" " "
	30.		

The Fixed Salts about as follows:

Chloride of Sodium,	7.3	Fixed Salts.	parts in 1,000 of Urine.
Phosphoric Acid,	2.2		" " "
Sulphuric Acid, -	1.8		" " "
Lime, - -	.3		" " "
Magnesia, - -	.2		" " "
Potassa, - -	2.		" " "
Soda, - - -	.2		" " "
	14.		

Be it remembered, that the above table represents the average proportion of ingredients in healthy urine, and is adopted as the "Standard." It is divided into three general parts, to-wit: Water, Organic Matters, and Fixed Salts, that the more minute and specific ingredients may be classified therein. By an acquaintance with the optical qualities merely, a single glance at the urine, will sometimes enable one to designate the probable course of deviation from the healthy standard, whether too aqueous or too solid, too light or too dense, etc., and from these significations, to determine to which of these divisions it belongs. Dr. Bowman says: "The urine passed during a diseased state of the system is almost invariably more or less altered in its composition, and frequently presents physical peculiarities, as of color, opacity, etc., which are at once apparent on the most cursory examination."

The slightest deviation from the appearance of the urine in health, is so apparent to the eye, that nearly every person has been a casual observer of the fact in his own case, when "not well."

How many of the unlettered, even, have learned to attach great importance to the change in the appearance of their urine, from having accidentally observed it perhaps when sick, describing it as being too pale or too red, too light or to dark, milk-white or muddy, etc.

Whenever the natural ingredients of urine maintain their proper relation to each other, that fluid will be of a clear but somewhat pale amber-color, something like sherry wine. Its transparency, however, will be slightly affected on cooling, by the gradual subsistence of a slight mucus cloud, derived from the bladder or perhaps the urinary passages. "Whenever, however," (says Bird,) "one or other of the ingredients exists in real or comparative excess, or a new substance is superadded, the urine does not generally remain clear, but either immediately on being voided, or at least upon cooling, becomes more or less turbid." "It is likewise considerably influenced by disease," says Markwick, "being particularly strong in color and odor, in all cases of fever and inflammation, for instance, and almost entirely wanting in both these characteristics in anemia, hysteria, etc., while in diabetes, immediately after the urine is voided it sometimes resembles whey in color, and subsequently, when fermentation commences, is of an alcoholic nature."

The color of the urine depends somewhat on the degree of concentration, and hence the the necessity of ascertaining its *specific gravity.* This is placed, as before remarked, at one thousand and twenty in males at adult age, as the "standard." This density is a little lessened in females of the same age, but not sufficiently in every case to distinguish the sex of the patient, so many other circumstances tend to produce the same difference in urine.

The urine passed in the morning, or after a night's rest, called by writers, *urina sanguinis,* as having come more from the whole of the blood, when the system was in a state of repose, and less likely to be influenced by exercise or excitement, is that which should always be preferred, as furnishing the fairest specimen of the average density, etc.

As the average specific gravity of healthy human blood is less in the female than in the male, so is the average specific gravity of the urine also less in the female. The female has been considered the "weaker vessel," and it would seem to be so in point of fact, as well as anatomically, in the structure of the human frame—physically in the formation of the muscles, chemically in the constituents of the blood, or mentally, as is considered by some, in the "power of the brain." The bone is more rounded in form, the muscle is finer in texture, and the blood is lighter and weaker, and the urine is lighter in color, and its average specific gravity is less.

Having set forth, in some measure, the general principles upon which the system of Uroscopia is founded, we will give a concise view of the mode of procedure in the clinical examination of a specimen of urine, and the determination of the nature of the disease.

On the presentation of a specimen of urine for examination, the *name* and *age* of the patient is obtained and recorded in a book, leaving sufficient space for a record of the quality of the urine, the symtoms and nature of the disease, and in practice, the prescription of medicine. The name of the patient serves for convenience of reference in after examinations, and the avoidance of confusion in subsequent prescriptions that may be required.

A knowledge of the age is necessary, inasmuch as even healthy urine varies in composition in children, adults and the very aged; it being more aqueous in children, more acrid and solid in adults, and more fetid, etc., in very old persons. And besides it serves the better to fix a *pro rata* "standard" of the urine, according to such age, and also to regulate the doses of medicine in accordance with the same, when medical aid is required.

We then place the urine in a test-tube, and take observation of its color and opacity, the deepness or paleness of the one, the lightness or heaviness of the other, which will alone many times enable us to determine very correctly the class, at least, to which such specimen belongs. Be that as it may, however, as yet, we then examine its specific gravity, by immersing therein the gravimeter or urinometer, which is a small glass instrument with two balls and a narrow stem, the stem being marked by degrees corresponding to the depths it will sink, according to the density of the fluid. This gives us a knowledge of its density or specifiic gravity, and in addition the assistance it renders us in the classification of the specimens, "puts us in possession," says Bird, "of the data necessary for the calculation of the proportion of solids excreted by the kidneys; and this not unfrequently enables the physician to detect a previously unsuspected cause of emaciation."

After the specifiic gravity of the urine is obtained, the acidity, alkalidity or neutrality of the same may be tested. This is done by immersing in that fluid blue and red litmus-paper. If the urine be acid, the blue color of the paper will be changed to red; if alkaline, the red will be changed to blue; but if no change occurs in either, the urine is neutral.

A knowledge of the state of the urine in this particular, is highly important in itself, but more especially in connection with subsequent chemical analyses of the deposits, as it forms a guide for the selection of tests, and the direction of experiments therein.

From the color, opacity, cloud, pellicle, or deposit; quality, odor, freshness or putridity; acidity, alkalinity or neutrality of the urine, a tolerably close conjecture at least, may be made as to the character of the fluid; which a few well-directed experiments will generally be sufficient to decide.

If the urine be very high-colored, blood may be suspected; if so, a dark coagulum will be found on boiling in a vial or test-tube.

If on boiling, the dark coagulum be not present, and on adding a little hydrochloric acid, a decided red color is formed, it is only an excess of coloring matter.

If there are any traces of a pink color present, the addition of a warm solution of urate of ammonia will throw down a pink-colored precipitate, called *purpurine.*

If the specific gravity be higher than one thousand and twenty-five, urea may be suspected to be in excess; if so, by the addition of an equal bulk of nitric acid, keeping the glass cool by allowing it, to float in water, a crop of crystals of nitrate of urea will, in a short time, appear.

If urea be not found, sugar may be suspected; when by mixing one-third part of liquor-potassa with the urine and boiling it quickly, for five minutes, the liquid will assume a brownish or bistre tint, if sugar be present.

If the urine be fully acid to test-paper, and a precipitate be formed on boiling, and by the addition of nitric acid the precipitate is dissolved, the *earthy phosphates are in excess.*

If the urine be alkaline to test-paper, and a coagulum is

formed on boiling, and by the addition of nitric acid the coagulum is not dissolved, *albumen is present.*

If the urine be alkaline and no coagulum be found on boiling, and a precipitate is thrown down by nitric acid, the production is *uric acid* in excess.

If *bile* be present in a given specimen of urine, it generally gives a more or less decided yellowish-brown color. To determine, pour a few drops of the urine on a white plate, and then carefully add a few drops of nitric acid; the liquid becomes successively pale-green, violet, pink and yellow, the color changing rapidly as the acid mixes with the urine. If the deposit of urine upon cooling, be a mere floculent, cloud-like substance, easily diffused again on agitation, not disappearing on the addition of nitric acid, it is chiefly made up of healthy mucus, epithelium, or in women, an admixture of leucorrhœal discharge.

"If the deposit is ropy, and apparently viscid, add a drop of nitric acid; if it wholly or partly dissolves, it is composed of phosphates, if but slightly affected, of mucus. If the deposit falls like a creamy layer to the bottom of the vessel, the supernatant urine coagulable by heat, it consists of pus."

"If urine, on cooling deposits a white sediment, it is urate of ammonia, phosphates or cystine. The first disappears on heating the urine, the second, on the addition of dilute nitric acid, while the third dissolves in ammonia, and the urine generally evolves an odor of sweet-brier.

"If the deposit be colored, it consists of red particles of blood, uric acid, or urate of ammonia, stained with purpurine; if the first, the urine becomes opake by heat; if the second, the deposit is in invisible crystals; if the third, the deposit is amorphous, and dissolves on heating the fluid."

The apparatus and reagents required for a superficial examination of the urine, are but few; in fact, nearly all the necessary tests and experiments can be conducted in a satisfactory manner, in a very short time, and with but few instruments or reagents:

A Urinal or two, holding two or three ounces.
A Gravimeter or Urinometer, made small.
Red and Blue Litmus-Paper.
A Test-Tube.
A Pipette.
A Spirit-Lamp.
A Blow-Pipe.
Two or three Glass Slips.
A Microscope or Object-Glass.
Acetic Acid.
Nitric Acid.
A Solution of Potassa.
A Solution of Hydrochloric Acid.
Alcohol, Ammonia and Ether.

In enumerating the foregoing conditions of the urine, and some of the changes produced by reagents, as also in the apparatus, we have drawn somewhat upon the already published works of Bird, Bowman, Griffith, etc., whose works would be found of most important service to the student as well as practitioner, as they have been a valuable guide to us in conducting our experiments, both in superficial as well as more complete investigations. And even where great exactness is required, as in cases of poison, etc., wherein the precise *extent* of the abnormal ingredient in the fluids is important to the case, these works will be found of great value to the profession.

The average proportion of the chief ingredients in the urine being reduced to a "standard," the quantity of water in a healthy specimen of that fluid, is placed at nine hundred and fifty-six parts in one thousand.

An excess over that quantity will be known by its more limpid-water appearance, while the reverse or a deficiency in the proportionate quantity of water, will be recognized by the "heaviness of body" which it will exhibit, because of the undue proportion of organic matters or fixed salts it must contain. The optical properties even thus noticed, will many times determine whether the specimen be above or below the standard specific gravity. The application of the gravimeter, however, will determine the precise course and extent of such deviation from the healthy standard, and will thereby suggest important points for consideration.

This excess of water in some diseases, as in hysteria, and some others, in which want of red globules is the characteristic, amounts to nine hundred and ninety-five parts in one thousand, while in some other diseases it has been known to be reduced to less than four hundred and eighty parts in one thousand, as in cholera, etc. The average quantity of *urea* contained in the standard of healthy urine being fourteen and five-tenths, an excess of that substance is characized by the "body" or weight which it adds to the urine.

Its specific gravity will be greatly increased. In some diseases wherein that substance reaches a very great excess, the urine has been known to contain more than thirty parts in one thousand. In these cases the urine looks almost like a moderately-strong solution of saltpetre, with a little coloring matter added. It is considered to be one of the products of the destructive assimilation of the tissues of the body. (See Chart, Fig. 1).

The average quantity of *uric acid* contained in the standard of healthy urine being four-tenths parts in one thousand, when in excess in that fluid, "it usually exhibits rather a higher color than the healthy secretion, either a deep amber or a reddish-brown." In some diseases, such as inflammatory fever, it amounts to two, or even three parts in one thousand of urine, in which case its sediment is formed in the bottom of the glass on cooling, of a decidedly red color. It indicates inflammatory diseases, sometimes eruptive, and in some cases calculous diseases.

The average quantity of *extractives* contained in the standard of healthy urine being fourteen parts in one thousand, an excess of these always gives to this secretion a higher color than natural. When the excess is very great, the urine becomes of a decidedly reddish color, and when cool sometimes deposits a quantity of brownish or even bluish-black sediment. This occurs generally in dissolved states of the blood, as in the last stages of typhoid fever, measles, etc. This kind of sediment is readily soluble in alcohol, a test whereby it may always be determined.

The average quantity of *mucus*, in healthy urine, according to the "standard," being two-tenths parts in one thousand, an excess of that substance, while it may produce no change in the color, yet it is very readily detected by the deposition of a very viscid, tenacious substance, consisting, according to Bowman, "chiefly of mucus mixed with epithelium, which, when agitated, does *not* mix again uniformly with the fluid, but coheres together in tenacious, ropy masses, entangling and retaining numerous bubbles of air." It is characteristic of inflammation of mucus membrane, bladder, urethra, vagina, etc., indicating, sometimes, gonorrhœa, leucorrhœa, etc.

The average quantity of *ammonia* given in the standard, being nine-tenths parts in one thousand of healthy urine, the secretion containing it in excess, "will most frequently be found high-colored, dense and turbid." It generally indicates idiopathic fevers, local inflammatory action, etc.

The rest of the ingredients composing healthy urine, as per "standard," being the "fixed salts," the whole comprising fourteen parts in one thousand of urine, are principally to be detected by chemical tests or reagents, or by the use of the microscope. "In the examinations of urinary deposits," says Bird, "the microscope will be found to afford the most valuable and ready assistance, the single microscopic inspection of a deposit often rendering its true nature at once apparent."

Chloride of Sodium, when in excess in urine, is very readily determined by the evaporation of a small quantity of the fluid containing it, on a slip of glass, when the salt will be found "in crystals shaped like daggers or crosslets." (Bird.) (See Chart, Fig. 2).

"The soluble phosphates," says Bird, "must be regarded as being derived directly from the food and from the blood, when in the act of being formed into muscle. The insoluble phosphates forming part of the structure of the body derived originally from the blood, are conveyed to the urine in the process of metamorphoses of tissue." Now, phosphorus enters largely into the composition of some of the structures of the body, especially in that of the brain and nervous system. In certain diseases of the brain, there has been found an actual deposition of this nervous matter in the urine, showing that the brain even follows the same general rule of other organs, and contributes its substance in disease, to the urinary secretion, and that such may be detected therein.

Oxalate of lime, urate of soda, and others of the "fixed salts" of the urine may be readily detected by chemical tests, or by microscopic exanination, for which, see Chart.

The presence of the "fixed salts" generally may be accounted for in the urine, by their presence in the food, drink, etc., from which they enter the circulation, the insoluble or unassimilated portions being separated by the kidneys. An excess of any of these may prove a source of irritation or irregularity in some part of the animal economy and is therefore worthy of notice; a substance called *kiestein* is found in the urine of pregnant women, and will demand attention; as, also, *spermatazoa*, found in the urine of persons laboring under seminal diseases; other substances, as *torulæ*, *vibroines*, etc., have all demanded attention, and will receive a notice in their respective places in this work.

INFLAMMATORY FEVER.

We come now to the practical part of our work, the application of our theory to particular cases in practice. And first, let me remind you again, that you will be much facilitated in all your examinations, by obtaining the *age* of your patient when he is not present; as by that you may be saved the trouble of testing the urine of very old persons, for worms, croup, or diseases peculiar to children alone, or of very young persons, for diseases peculiar to middle-life or old age.

In the above disease, to-wit: inflammatory fever, we know that it generally seizes young persons, in the flower of their age, and when full of blood. But as inflammatory rheumatism is very nearly allied to it in point of general conditions, both of blood and urine, and as this latter disease sometimes attacks the more aged, the necessity of keeping in view the age of the patient will be apparent.

The chief local symptoms of inflammation are defined to be *redness*, *heat*, *pain* and *swelling*. "The redness of an inflamed part is obviously due," says Williams "to the increased quantity of blood in the vessels. The heat is dependent on the increased flow of blood through the part, "and may be considered the representative of the amount of determination of blood." The swelling of an inflamed part is caused by the enlargement of the vessels, the consequence of the increased flow of blood to the part; while the pain is caused by the exaltation of sensibility in the nerves, and the tension or pressure arising from the swelling. The irritation of inflammation thus produced, frequently extends itself to the whole body; and this is called inflammatory fever.

"Among the most important general effects of inflammation," says Williams, "must be noticed the change in the condition of the whole of the blood." An excess of fibrin exists in all true inflammatory diseases, especially those of a sthenic character;" in some cases M. M. Andral and Gavarret found the proportion of fibrin in the blood, to be as high as twelve parts in one thousand, the average in health being three. Even when inflammation supervenes in the course of another disease, "there is always an augmentation in the quanty of fibrin in the blood."

The red globules of the blood have also been found largely increased in some inflammatory diseases, especially those of a sthenic character also, rising in some instances to one hundred and eighty-five in one thousand, the "standard" in health being one hundred and twenty-seven parts.

"The most sensible influence," says Prof. Jones, "which inflammation exerts upon the blood, as determined by analysis, is manifested in the increase of its fibrin. This, it is said, is uniformly augmented in quantity during the progress of inflammation, commencing from the first establishment of this abnormal excitement, and diminishing as it declines." "These facts," he says, "are established by the very best authorities."

Now, in view of this altered state of the blood in inflammatory fever, as represented by the authorities herein quoted, the quantity of fibrin always rising above the average, and sometimes amounting to three hundred per cent. above the standard in health; and in view of a corresponding increase of the red globules of the blood in every case of this kind of fever; and knowing that the secretion of urine proceeds entirely from the blood, you will readily anticipate the characteristic indications to be found in that fluid under these circumstances. But let us see what are the facts in the case, and what has been observed by eminent medical authors on the subject.

"In inflammatory affections," says Simon, "and in those diseases which are accompanied by that form of fever which is termed sthenic or synochal, the urine differs greatly in its properties from normal urine." He does not refer the probable cause of the changed condition of the blood in these diseases, to the diseased organ alone, but in part to the reaction which manifests itself throughout the vascular system. He says, "if the change in the constitution of the blood bears an accurate and inseparable relation to the fever, there can be no doubt that the change in the constitution of the urine must be in the relation to the same cause, for the urine is separated from the blood, and was previously an integral constituent of it, and because, farther, every alteration in the constitution of the blood, must involve corresponding changes in the secretions and excretions, and more especially in the urine. Since like effects follow like causes, and since in inflammatory affections the vascular system similarly participates in the disturbance, we may assume *a priori*, that similar changes will occur in the urine—a point confirmed by experience."

The urine discharged during inflammations is generally termed febrile urine, but there is an objection to this term, here. Since the cause of the change in the urine will be found in the character of the morbid change going on in the system, and will be inflammatory or febrile, according to the character of the disease, whether it be synochal, idiopathic, etc.

In order to take a correct view of the composition of the urine then, we must bear in mind the composition of the blood, and the reaction of the vascular system, and authors tell us that if in these inflammatory cases, much blood be abstracted, the quantity of fibrin in the blood is lessened and the urine becomes immediately changed, being clear, specifically lighter, and the amount of urea also decreases—absolutely and relatively.

The following is given as some of the general characteristics of the urine in inflammatory affections, by Simon :

"The urine is darker than usual, and is of a yellow, brown, or reddish-brown tint; it has an acid reaction, and is generally of high specific gravity. With respect to its most important constituents, the urea is either absolutely increased, or is at the ordinary physiological average. The uric acid is always absolutely increased, and so are the extractive matters. The salts are always absolutely diminished, especially the chloride of sodium."

Prof. Eberle says that "revolution in this form of fever is almost invariably attended by general and free urination, etc.," and that "a reddish or pale sediment in the urine is a never-failing concomitant in the crisis of this fever."

Prof. Beach says that "at first the urine is *very high-colored*," etc., "and when terminating by revolution a never-failing concomitant is found in a reddish or pale sediment in the urine."

"The more acute and fixed the inflammation," says Williams, "and the smarter the fever, the more abundant is the deposit, and the more free is the patient from disease afterward."

"The lateritious sediment in the urine, which is constant on the decline of inflammatory fever, is a pretty certain symptom of its subsidence, or of the amelioration, at least, of the inflammation."

This excessive deposit, on the subsidence of the inflammation, indicates an increased exertion of the solid constituents of the urine, some of which had probably been delayed in their exit during the first febrile excitement, because of the scanty secretion of that fluid.

During the first stages of the inflammatory excitement, "the urine is marked by a more bright and highly-colored appearance, together with a reddish-tinged floculi, which seems to have been *wrung*, as it were, from the excited organs," says Eberle.

"That these are the excrementitious matters that had accumulated in the blood, when the skin was dry and hot and the natural secretions were diminished, as in the first stage of this fever, we can scarcely avoid to conclude," says Williams.

That these are the decayed tissues which are always taking place in such cases, there can be no doubt, together with the particles that are in excess in the blood.

The blood, in this disease, being characterized by an excess of fibrin, as also an excess of red globules, how natural that wé should find in the excrementitious matters which proceed therefrom, a fluid not only of a "more highly-colored appearance, but "tinged with a reddish floculi," the real fibrin itself, which has been wrung from the excited organs, imparted to the blood, and separated therefrom by the kidneys.

When the crisis has formed—when the great change has come over the patient, a great change seems to come over the urine also—"the pale-reddish sediment, the never-failing concomitant," etc. This sediment, however, true to the never-varying rule, is composed of the same substance as that before, metamorphosed only; together, perhaps, sometimes, with disintegrated parts from other tissues, as the chemical analyses of Bird and others have fully shown.

These organic matters, red particles, floculi, sediment, etc., which have been in such excess in the blood, and carried out by the urine, are metamorphosed into uric acid, which chemical analysis and microscopic observation discovers to be in the urine in great excess, "and commensurate with which excess, will have been the activity or severity of the inflammation." "For, in all acute inflammatory disorders," says Bird, "a considerable increase in the quantity of uric acid will occur, and deposits of this substance, either free or combined, will appear in the urine," "generally amounting to more than double the average," says Becquerel.

The urine then, in this disease, from the forming stage to the crisis, will be found bright and high-colored, amounting even to a cherry red, with a reddish-tinged floculi suspended in it, as may be observed by holding it in a clear glass vessel before a light, the extent of color and amount of floculi varying with the extent of the disease.

From the crisis to the completion of convalescence, it is found to deposit crystals of every shade and tint, the deeper the color and the more extensive the deposits, the more extensive it shows the disease to have been, or the more quickly complete convalescence will ensue.

But should the patient continue to grow worse, until a dissolution of the blood takes place, the urine will gradually become more and more heavily loaded with these deposits, and will become heavier in body and darker in color, even muddy-like, or almost a coffee-ground appearance, which is the evidence of approaching death.

"When urine contains this acid in excess, it generally lets fall crystals on cooling; and they are sometimes sufficiently large to allow their figure to be defined without the aid of the microscope. Every shade of tint, from the palest fawn-color to the deepest amber or orange-red, is to be found in the deposit; the deeper the color of the urine, the darker the color of the sediments." (Bird).

It always reddens litmus-paper, and its specific gravity is above the average. When the urine is high-colored, the addition of a few drops of nitric acid will cause a deposition of an abundance of these crystals.

If you wish to examine with the microscope, Dr. Bird has given the most convenient plan, which he describes thus:

"Allow the urine to repose for a short time in a tall vessel, decant the greater portion, and pour a teaspoonful of the lowest turbid layer into a watch-glass; gently warm it to dissolve any urate of ammonia and to aid the deposition. Remove the supernatant urine with a pipette, and replace it with a few drops of water, then place the watch-glass under the microscope and the crystals covered by the water will become most beautifully distinct."

We have been thus elaborate in our explanations of the condition of the blood in this disease, and the different characters of the urine in the different stages, because it is the ruling one of a large *class* of disorders, some of the characteristics of which will be found throughout the whole list of inflammatory diseases.

From the foregoing history of the condition of the blood, and the peculiar character of the urine in this disease, it is plain that the treatment which would be the most effectual, or at least of considerable aid, would be such remedies as operate in diminishing the red particles, and reducing the excess of the fibrin in the blood.

Here lies the great practical value of our system: to arrive at the true cause of a departure from health, that the removal of that cause may be effected. The treatment which has been found most effectual in this disease, makes the correctness of our conclusions the more final. The various saline medicines, such as nitrate of potassa, and the alkalies, combined with vegetable acids, seem to have a good effect, by augmenting the elimination from the blood to the kidneys.

"These," says Williams, "being more or less diuretic, by supplying an alkaline base, unite with the acids, formed in the blood, and facilitate the separation of the matters by the kidneys."

Among inflammatory diseases, in which some of the general characteristics of the blood, mentioned under this head, will be found to exert a partial influence over the urine, in the manner here described, may be mentioned inflammation of the brain, lungs, heart, liver, stomach, bowels, kidneys and womb, also, acute rheumatism, etc.

But in all these, each have their other distinguishing marks, whereby they are known one from the other, generally because of the additional change in the constituents of the blood, its corresponding change being exhibited in the urine.

The urine, then, in the *advancing* stage of inflammatory fever, will be of specific gravity above the average; it will redden litmus-paper; will be highly-colored, sometimes cherry-red; and will contain a reddish-tinged floculi.

In the *declining stage* it will first deposit a pale-reddish sediment, becoming more highly colored, even to a deep orange-red; finally, let fall an abundance of uric acid crystals. (See Chart, Figs. 3 and 4).

In the *last stage* it will be dark-colored, thick, heavy and muddy-like; sometimes like coffee-grounds.

INFLAMMATION OF THE BRAIN.

FROM analyses of the brain, made by different authors, it appears that the medulla oblongata and spinal cord contain the same constituents, a preponderance being given to the brain only, of phosphorus; and that the amount of cerebral energy is proportioned to the relative amount of the phosphoric element in that organ. From the table of analyses of the brain of infants, youths, adults, aged persons and idiots, drawn by L'Heretier, from his own researches, the amount of phosphorus in the brain of infants and idiots was scarcely appreciable, rising in youths, to a maximum in adults, and declining again in the aged. In the brain of a complete idiot there was not a trace of phosphorus to be found.

"In inflammatory affections of the brain, and also in those of the spinal cord, especially in chronic cases," says Simon, "the kidneys and bladder sympathize to a high degree."

The urine, in these cases, is either only slightly acid at first, if at all, is generally neutral, and, in a very short time, becomes alkaline.

"When first discharged, the urine is clear," says the same author, "generally of a bright yellow color, and possesses rather an unpleasant odor. If allowed to stand, a glistening pellicle often forms quickly on the surface, consisting partly of crystals of ammoniaco-magnesian phosphate, and partly of amorphous phosphate of lime, as may be seen by the microscope." After a time, the urine becomes turbid, and deposits a sediment of earthy phosphates and mucus, and a strong ammoniacal odor may be given off, when, upon the addition of hydrochloric acid to the urine, a well-marked effervescence will be produced by the liberation of carbonic acid."

A case is related in which a man, aged forty years, was brought into a hospital with a severe cerebral affection; he soon sank into a state of deep coma, and the urine was emitted involuntarily. It had an ammoniacal odor, an alkaline reaction, and soon deposited a sediment of mucus and earthy phosphates, and, upon the addition of nitric acid, a brisk effervescence took place.

This characteristic of the urine is to be found in many cases of insanity, and sometimes in meningitis, and in encephalitis of children even. But "inflammatory affections of the brain and spinal cord are not the only diseases," says Simon, "in which carbonate of ammonia is formed in the urine, as we shall subsequently show that alkaline urine is frequently observed in diseases of the kidneys and bladder, and in nervous fevers." We will remind him, however, that even in these, the pathological condition of the brain may have much to do with it.

The symptoms of inflammation, as mentioned in preceding pages—redness, heat, pain and swelling—would be those of the brain, in inflammation of that organ; the consequence of an increased quantity of blood in, and flowing through the vessels, producing enlargement and consequent pressure upon that organ itself. Hence, the same conditions of blood as are presented in general inflammation, will be partially met with in this, together with the addition of a peculiar nervous substance, the consequence of a chemical change in the substance of the brain, "consisting," says Prof. Buchanan, "of a disintegration of the nervous matter from that organ."

Physiologists have considered the functions of the brain so dependent upon the momentum of the blood in its vessels, that, "where the neck is very long, and the brain far removed from the heart," says Burrows, "the faculties are more limited and its functions less active, whereas, a short neck and approximation of the brain to the heart, usually coincides with cerebral energy."

Now, as phosphorus enters more largely into the composition of the brain and nervous system, but more especially the brain, than any other organ or part of the body, as the disintegration of that organ is going on by inflammatory action, these disintegrated particles are being taken up by the blood, and conveyed into the urine, by which it is eliminated; metamorphosed, however, into phosphoric acid and other phosphates, perhaps.

The urine, in the disease now under consideration, as a general rule, will be paler than natural, although marked with a reddish-tinged floculi or red particles of inflammation. Its specific gravity will be a little above the average, and constantly alkaline to test-paper. The deposit, which falls to the bottom, on repose, will be of a whitish, or sometimes of a reddish-gray color. This consists of phosphates, and sometimes is exhibited in excess of phosphoric acid, amounting to as high as eight parts in one thousand of urine, two and two-tenths being the average. (See Chart, Fig. 7, also page 62).

To test this deposit, by chemical action, you have but to add a little dilute hydrochloric acid to the specimen, and it will immediately dissolve it, leaving the urine more clear; while liquor-ammonia or potassa produces no effect.

When, therefore, you find the urine paler than natural, although it may be somewhat hazy, and a little tinged with red, specific gravity a little above the average, alkaline, deposit of the above-mentioned kind and color, having tested it chemically, and viewed the deposit under the microscope, and found it to correspond in appearance to the "prepared specimen" or the figure in the Chart, you will be enabled to pronounce at once upon the nature of the disease. These characteristics of the urine if not, when taken separately, certainly when combined, will enable one to discover not only the brain to be the seat of inflammation, but to partially estimate the extent also of the organic mischief going on at the time, by the amount of disintegrated particles imparted to that fluid, from the metamorphosed substance of that organ.

"The occurrence of these phosphates in the urine," says Bird, "must be regarded as of serious importance, always indicating the existence of important functional, and too frequently, of even organic mischief."

If this disease becomes of a dangerous and aggravated character, and the depressed state of the cerebral and nervous energy becomes extreme, "an extensive elimination of *urea* will be discovered in the urine," a fact necessary to be borne in mind in the examinations in the last stage of this disease.

Prof. Buchanan says of the brain, that the chemical changes of its substance consists of the disintegration of the nervous matter, and that these changes are accompanied by corresponding changes in the secretions, and that the kidneys, of all organs, appear to sustain the most intimate relations thereto.

In the *advancing stage*, the specific gravity of the urine will be above the healthy standard; it will be alkaline to test-paper; and paler than natural, but hazy, and tinged with red.

In the *declining stage*, the urine will be heavier, and the deposits whitish or reddish-tinged, and sometimes grayish sediment, and will have an excess of phosphoric acid.

In the *last stage*, there will be deposits of urea.

The indications of cure are :

1st. To equalize the circulation by sinapisms and pediluvia, with quickly-evaporating lotions to the head, as ether, alcohol, and sometimes warm water.

2d. Moderate the circulation, to remove the tension, by sedatives, such as gelseminum, veratrum, viride, etc.

3d. Remove the offending cause from the blood by diuretics of the vegetable character, as uva ursi, digitalis, buchu, etc., all three of which indications may be conducted at the same time.

INFLAMMATION OF THE EAR.

THIS is a peculiarly painful disease, and one the character of which is not always so readily determined without the closest scrutiny. Men have been treated for inflammation of the brain, who merely labored under inflammation of the ear, and upon the other hand patients have frequently been treated for "nervous otalgia" when all the pain and dangers of true inflammation existed, for it must be remembered that the sense of *hearing* is not only many times destroyed by neglect or improper treatment of deafness, but that extension of the inflammation to other parts: the bone, membranes of brain, enecphalon etc., may prove dangerous to life.

It is in determining between "nervous otalgia," a *neuralgic* affection and true inflammation of the ear, that the urinary secretion becomes available, and is, more than all the other signs, to be depended on; the difference in the general characters of the urine of inflammation and of nervous urine, being the distinguishing marks.

A case of the above disease came under our own observation a short time ago, the inflammatory action of which was only to be correctly determined by the appearance and quality of the *urine*, the sympathetic action of the nervous system having almost completely masked its true nature. The great nervous irritability of the patient led the attending physician astray, nor did he discover the mistake until his attention was directed to the continually high-colored and inflammatory appearance of the urine.

So intimately connected with disease is the urinary secretion, that even, in "severe dysuria," says "Wilde, urine has been discharged from the ears." The urine in otitis is always scanty and high-colored, bearing all the general marks of inflammation, "and toward the termination of the acute symptoms, deposits a copious pinkish sediment." (Wm. R. Wilde, on Diseases of the Ear).

The above-quoted opinions correspond with our own experience practically, having suffered the disease, as well as experimentally in the examination of patients.

We were suffering violently with otitis, the consequence of a cold, bleak ride along the bank of a river, with the piercing wind howling around our head, until our ears were well-nigh frozen. The inflammation being violent, we suffered intense pain for some time and passed several restless days and sleepless nights without abatement, yet it was only "ear-ache," and our friends thought it was "nervousness," which was not at all flattering to our character nor agreeable to our own opinion—more especially as we had observed the well-makred characteristics of inflammation in the urine.

In a little while longer this opinion was fully confirmed by the termination of the inflammatory action in suppuration, and purulent discharge from both ears, and a subsidence of the pain, as also the irritability of the nervous system.

The character of the urine, whether inflammatory or nervous, is always too well marked to deceive the merest tyro who will observe it.

INFLAMMATION OF THE LUNGS.

The blood, on exposure to the air in the lungs, is seen to change from purple to florid. During respiration the air exhaled contains less oxygen, and more carbonic acid than that inhaled; the blood becoming so far chemically changed that it will now contain more oygen and less carbonic acid. The production of this chemical change seems to be one of the principal offices the lungs have to perform.

Of all excretions, that of carbonic acid seems the most essential to the continuance of life, and any sudden interruption to its progress proves immediately fatal. We have many opportunities of recognizing the influence of carbonic acid on animal life as that of a poison, positively; and we can trace all degrees of this agency from the inconvenience of breathing the ill-ventilated atmosphere of a close room, to the rapid extinction of life by inhalation of the vapors of burning charcoal.

Whenever the extent of respiratory surface is diminished by any organic disease of the lungs, or the access of fresh air through them is obstructed, defective exhalation of carbonic acid attends, and its accumulation in the bood results.

"The most fruitful source of inflammation," says Calkins, "is a depraved state of the blood; and owing to some chemical or other affinity on the part of different ingredients in it, for different tissues or organs, the localities of the inflammation created are varied according to the nature of the causes."

In acute inflammation of the lungs, the blood soon becomes surcharged with carbonic acid, because of its non-elimination, the swelling and consequent pressure in these organs preventing that freedom of action necessary to the full performance of the offices expressly devolved on them.

The painful stitches which cause the patient to cry out in distress in this disease, arises, in part, from the irritating influence of this carbonic acid, and the increased volume of blood to the organ, or its consequent tension.

The substance of the lungs tremselves, chemically considered, contains a greater proportion of carbon than that of any other part of the human system; hence, not only does the imperfect elimination of carbonic acid from the blood become a source of disease, by its retention in the circulation, but when a disintegration of the substance of the lungs takes place, and these metamorphosed particles are taken up by the blood, an additional source of excitement is manifested therein.

Now this double source of disease, as it were, together with the extreme delicacy of the tissues of these organs, and the constant requirement of labor devolving upon them, is the main reason that simple inflammation of the lungs is so prone to run into confirmed consumption, or chronic disorganization of these tissues.

The blood, in inflammation of the lungs, usually exhibits the general character of phlogosis, more decidedly than in other inflammatory diseases. The clot is said to be rather below the ordinary size, very consistent, and does not break down for a considerable time; its surface is covered with a buffy coat, and is more or less cupped; the serum is of a pale-yellow color, and the quantity of solid constituents is usually less than in healthy blood. The maximum of fibrin is said to be much larger than is ever discovered in inflamed blood, while the hemato-globulin is very far below the amount in healthy blood.

"The urine in pneumonia," says Simon, "is subject to considerable variations, dependent upon the extent of the disease, and the degree of inflammation. In severe inflammations the urine is very dark, of high specific gravity, and frequently sedimentary, especially at critical periods and during the fever."

Becquerel, however, once found that the urine deposited a sediment on the day when the fever ceased. He also found the urine to remain acid during the whole period of inflammation, as also, during the period of convalescence. The small quantity of solid constituents in the blood in this disease, coincides with the loaded state of the urine.

The urine, in this disease, is "scanty and high-colored," most authors contenting themselves with this stereotyped assertion alone, but other and more definite characters are certain to be observed in this stage of that dangerous disease, and difficult one to cure.

The color of the urine will be found of a brighter red than natural, not brownish nor dingy, but bright, light crimson-tinged or cherry-reddish color; and in the forming stage even, it will always be marked by the bubbles of carbonic acid, commonly called a "bead," which hold for some considerable time therein, if the vessel containing the urine be kept closed.

Dr. G. Bird, on describing a certain kind of deposit in the urine, says that "it is so constantly an attendant on the slightest interference with the cutaneous transpiration, that a common cold is popularly diagnosticated whenever this state of things exist;" and we think that we may here venture the assertion that this "bead" is *universally* present on every accession of what is termed "taking cold," even though it be during the progress of other diseases.

The specific gravity of the urine in this disease, is always above the average in health, being loaded as it is with matters, the product of non-assimilation of food, together with the disintegrated particles from the inflamed organs.

By the addition of a few drops of nitric acid to half an ounce of a recent specimen of the urine, a thick whitish coagulum, or collection of a light, fine down-like substance, will form in a table across the middle of the glass, leaving the upper and lower portions more clear. This distinct table, however, is more or less heavy, and sooner or later extends to the whole, according to the quantity of said substance, which amount is the measure to the extent of the inflammation. When this fine, white, down-like substance extends to the whole of the urine in the glass, it is rendered quite opake, or nearly so, and upon farther repose it deposits the same, transformed into mixed phosphates and other amorphous substances.

The urine in this disease is always acid to test-paper, therefore, when you find it containing the bright-reddish tinge of inflammation also, carrying a "bead" representing bubbles of carbonic acid, conglomeration of substances forming a table midway across the vessel, on the addition of nitric acid, and finally settling at the bottom, after first extending to the whole of the urine in the glass, and exhibiting under the microscope amorphous substances, or mixed phosphates, of lime, soda magnesia, etc., with perhaps traces of mucus or albumen; inflammation of the lungs may be confidently expressed and decided upon.

If this substance, which is thus loading down the urine, as it were, should happily be carried off with sufficient rapidity at the onset of the disease to relieve the lungs of the tension, and the blood of its overload of substances, a favorable prognosis may soon be made. This is best done by diuretic salts, in combination with diaphoretic teas and other medicines that increase the cutaneous transpiration, also; for by this, as well as the urinary secretion, may the office of the lungs be fulfilled for a time. In connection with these counter-irritants over the region of the lungs, sinapisms, etc., will be of material service, in the first stage particularly, in removing the swelling and tension thereof.

After the acute inflammatory action is removed a more tonic and nutrient medication may be profitably employed, say, hydrastine, iron, wine, etc., for the supply of the wear and tear, the blood and system has undergone.

But, should the functions of the depurative organs or kidneys be inadequate to the task, and the transpiratory functions render no assistance to the lungs, in this critical moment, the substance or tissues of these organs themselves are likely to give way or yield to the overwhelming power, when the more-to-be-dreaded disease, consumption, will take the sway.

In such cases the specific gravity of the urine is generally above the average; it is acid to test-paper; its color-bright-red, or crimson-tinged; presents carbonic-acid bubbles, or "beads" on the top; a down-like table, forms in the center, on adding nitric acid, which becomes opake on further repose; and it deposits, amorphous, mixed phosphates, etc.

With these characteristics presented in a specimen of urine, inflammation of the lungs may be safely set down as the disease. No mistake will ever occur, if the "signs" are properly understood.

INFLAMMATION OF THE HEART.

In this disease, if there is no other derangement of the human system in connection, the condition of the blood will be but little altered from the natural, and consequently there will be but a slight change or alteration in the appearance or quality of the urinary secretion; hence you may be the more readily deceived in this affection, and more sadly disappointed, than in any other disease.

Failing to discover much difference of appearance, from natural healthy urine, or to detect any material abnormality in quantity or quality, or any great degree of excess or deficiency of any of the chief constituents of the urine of health, from your inspection of that fluid, you might readily, and would be liable to regard your patient as being very little, if at all, afflicted; when, in fact, there would be very serious danger to be apprehended, the patient feeling all the symptoms of inflammation of a most vital organ.

Although we have said the blood itself is very little altered in its general properties in this disease, still, there is enough change therein, to make the impression upon the urinary secretion sufficiently visible, when the character of that change is properly understood.

The heart being the propelling power of the blood, does not prevent it from receiving impressions from external causes, the same as other organs, it being subject to the same laws, and dependent upon the same source (the blood) for its nourishment, and for the removal of its decayed tissues, must rely upon the same channels of exit. The color of the substance of the heart itself is unlike that of any other organ or tissue of the body, being rather of a light mahogany-brown, and an excess of particles from this part, would, after the disintegrating process of inflammation, be imparted to the urine. This fluid therefore, would not only be hightened in color by the inflammatory action, which would give it a reddish tinge, but the peculiar light mahogany-brown color of the disintegrated particles which had been wrung from the substance of the heart itself, would be added thereto.

Now, if you can but imagine the color such a combination would form, you will have some conception of the appearance of the urine in this disease. We can only describe it as being more of a mahogany-brown color than the urine of any other disease, and to maintain the same appearance without change when exposed to the air, longer; and to be less easily affected by chemical tests, than is morbid urine generally.

Lecanu analyzed the blood of three men and five women who were suffering from this disease. Unfortunately he made no observations on the physical characters of the blood, and the quantity of fibrin was not ascertained; but the proportion of solid residue, and especially of the corpuscles, indicated a high degree of inflammatory action, and we have no doubt there would have been found an excess of the fibrinous element. The most remarkable feature seemed to be the extreme diminution of albumen in the blood.

Dr. Simon examined the urine in a case of very acute pericarditis. It was clear, of a deep fiery-red color, had an acid reaction, was of high specific gravity, and on being heated gave indications of the presence of albumen.

Zimmerman found *fibrin* in the urine of a patient with "endo-carditis of the right ventricle at the period of the commencement of hypertrophy."

These are about the whole of the different variaitions from healthy urine, or about all the characters we can enumerate as being peculiar to this disease alone, and they require to be scrutinized closely to be detected.

This disease soon involves other organs, when other characters will be given to the urine.

Treatment: Purgatives, sedatives, diuretics, with counter-irritants over the region of the heart, are the principal indications of cure.

INFLAMMATION OF THE LIVER.

In inflammation of this organ, we need hardly say that the usual symptoms of general inflammation, to-wit: "the tinge of red particles," etc., will be partially presented in the urine, but proceed to give the other indications manifested in that fluid.

First, however, let us see what we should expect in this disease. The liver is not only different in color and chemical composition from the brain, heart, lungs, or any other organ of the body, but its office in the animal economy is also different and peculiar. It is a depurating organ; its office being to free the blood, as it were, of a great portion of its foulness; this only gives it a more primary importance in its connection with many other diseases.

It is of a dull-red, or brownish color, and is composed principally of glandular substance, and is designed for the manufacture or secretion of bile.

The natural color of the liver being a dull-red, in inflammation, this color would therefore be increased; hence, any disintegrated particles therefrom would be highly or heavily colored indeed.

While the bulk of this organ would be enlarged by the inflammation, and its healthy secretion of bile interfered with or suspended, as it would be, some of the elements at least of this suspended secretion (bile) would be retained or find their way into the blood; and it is quite evident that such is the case, from the yellow color and jaundiced hue of the skin, eyes, and secretions in every case of biliary derangement. Hence, "in the elimination of that fluid, which is secreted from the blood," during this excitement, and which it is destined to rid the blood of this surplus or unhealthy matter, as also to remove the decayed tissues of the liver itself, with which the blood has become burdened during this inflammatory action; in view of the office this eliminatory fluid is required to perform, have we not good reason to expect to find it not only altered in its general properties, and changed from its natural appearance, but even partaking of the very qualities thus imparted to it by these circumstances? Most certainly we have; and, also, the ratio of difference between the average proportion of the chief constituents of a standard quality of healthy urine, and the urine thus discharged, to be the ratio of the disease, or the extent of the deviation from health.

In accordance with the philosophy above, in inflammation of the liver, true to the facts therein stated, the urine is discovered to be more highly or dark-colored, sometimes of a deep-saffron, or even a deep-green color, and in extreme cases amounting to a dark-brown color, according to the extent or severity of the inflammation.

Dr. Beach says, "the urine is scanty and high-colored."

Hooper says, "the urine is small in quantity, and sometimes saffron-colored, and if the disease continues for some days, it becomes very dark-colored, and deposits a red sand and a ropy mucus."

Dr. Newton says, "there is a jaundiced discoloration of the eyes, skin and *urine*."

In regard to the constituents of the blood, in hepatitis, it seems there has not been very accurate analyses made, but enough is known to satisfy us that the blood is not only far from being normal, but that the abnormality of the constituents are plainly visible.

"Nasse has occasionally seen it so highly colored with biliphein as immediately to tinge paper on being dipped in it." And Lauer has observed that a yellow-colored sediment is deposited by the serum upon the buffy coat, during this disease.

According to Coindet, "the urine in inflammation of the liver, instead of urea, contains a substance resembling bilin at least." In the analyzation of the urine of a man who was suffering from acute hepatitis, as conducted by Simon, the urine was scanty, had an acid reaction, was of a dark, reddish-brown color; on the addition of nitric acid, the brown color of the uriue changed into a decided green, and contained biliphein.

Becquerel analyzed the urine of a man attacked with icterus; the urine was very bilious, and deposited a yellowish sediment of uric acid.

Schonlein states that the urine, in hepatitis, is of a dark-red color, approaching a brown; that it usually contains biliphein, and that at the crisis, a rose-colored precipitate is sometimes formed.

Dr. Shearman says, "when the functions of the liver are deranged, and when bile is circulating in the system, the urine is very brown, and easily shows bile by the proper tests.

But why multiply testimony to prove an admitted fact? From the foregoing explanations, and descriptions of the urine in the inflammation in question, there would seem to be no difficulty in determining the nature of a case by a mere inspection of the urinary secretion alone.

The dark, reddish-brown, the deep saffron, the dark-green, or in severe cases, the very dark-brown colors, together with the red sand-like deposits, and the ropy mucus, according to the severity of the disease, will at once suggest to the mind the probable substance with which this fluid is loaded. If bile be suspected, as it most likely will be, the simple test of nitric acid, added to the same quantity of urine, on a white plate, the rapidly-ending play of colors, as before suggested, will be seen, as the acid mixes with the urine.

This disease, if not fully removed, is apt to leave such traces after convalescence, as lead to a chronic affection of that organ, which see.

Treatment: Small doses of podophyllin in combination with two or three-grain doses of leptandrin, Dover's powder, dandelion-root tea, with irritating plaster over the right side.

INFLAMMATION OF STOMACH.

As the stomach is the receptacle of all the nutriment that is required to keep up the wear and tear of the whole system, and as it is from this organ the blood receives the pabulum, which it so readily distributes to each part, according to its wants: so also, is it the store-house from which is received into the circulation, much of the deleterious substance that contaminates the body.

How quickly some substances enter the blood after being received into the stomach, is evidenced in cases of poisoning. Dr. Kramer discovered prussic acid in the blood of an animal which died from it in *thirty-six* seconds. Arsenic may be detected, both in the blood and urine, in a very short time after it is taken into the stomach. Orfila has found that arsenic most readily passes off during life by the urine.

In inflammation of the stomach, from whatever cause however, its regular functions of preparing nutriment for the blood, is interfered with, or entirely suspended sometimes, which interference or suspension will very suddenly change the character and composition of the blood, and therefore, morbidly affect the urine, according to the disorder of the digestive organs.

"Owing to the morbid state of the stomach," says Bennet, "the chyle is imperfectly elaborated, unfitted for the purposes of assimilation and nutrition, and on its being absorbed by the lymphatics and passing into the blood, the kidneys eliminate and throw out the effete matter in the shape of urate of ammonia, triple phosphates, or oxalate of lime, etc." "The state of the urine," says the same author, "is often a much more delicate test of the integrity of the functions of digestion, under all circumstances, than the other symptoms which we have enumerated." "Indeed we are surprised," continues he, "that so little attention should have hitherto been paid to this secretion" in disease of the stomach. The urine in this disease is highly colored, amounting, in some cases, to a deep or dark-brown, and if not turbid when first passed, will soon become so.

After it becomes turbid it soon throws down a dark-brown or dirty pink-colored sediment, which, upon examination, will be found to be urate of ammonia. If the disease progresses, phosphate of lime, or oxalate of lime will be found in the urine, which Bennet says, "when present, it often forms an iridescent film on the surface, like that which is seen on lime-water when exposed to air."

We have often noticed this last-mentioned appearance in the urine of inflammation of the stomach and bowels, as also epithelial scales, which are ever present in both these diseases. "The exfoliation of epithelial scales is sometimes so great, says Bennet, "that they are plainly visible to the naked eye," and so constant is their presence in urine from inflammation of the stomach and bowels, that we have frequently correctly diagnosed these diseases from this appearance alone." There is a slight difference only between these scales from the stomach, and such as are cast off during inflammation of the bowels; the former being flat and but little curled or curved. When the naked eye is unable to detect the difference, the microscope should be applied to.

The urine is always of high specific gravity alkaline to test-paper. It deposits urate of ammonia, triple phospates or oxalate of lime, and epithelial scales, bran-like and flat.

Treatment: Beach's Neutralizing Cordial, and mustard plaster over the region of stomach, and mucilaginous drinks.

This disease if not properly cured, or fully removed by a correct course of medication, is likely, after the most urgent symptoms have abated, to terminate in dyspepsia, which see.

INFLAMMATION OF THE BOWELS.

In this disease the internal coat of the intestines is the part affected, and as this is but a continuation of the same mucus surface as the stomach, the difference in the urinary secretion, from that described in the preceding disease, can not be very great.

The primary operations of digestion, however, is not so fully interfered with in this disease as in inflammation of the stomach; therefore, a measurable difference, in point of the depurating process, must occur.

It is a well-known fact that in inflammation of the bowels, dysentery, etc., although there is great thirst, the patient continually calling for water, there seems to be a very scanty supply of urine; in fact, an almost entire suspension of the urinary secretion takes place in every violent case. Indeed, when the symptoms run very high, as in cholera, there is a *total* suspension of that function, and a return of the urinary discharge only when the patient begins to convalesce. So universally is this the case, that a favorable prognosis may always be given in these diseases, when there has been a return of a plentiful flow of urine, especially if it be of a peculiar sedimentary kind containing the characteristics of epithelial scales.

The urine in this disease, will be characterized by the general high color and reddish-tinged floculi of inflammation, bordering, sometimes, on the dark-brown, and before it becomes turbid a large quantity of somewhat-curled, bran-like epithelial scales, each having a convex and concave side, may be readily seen thickly interspersed through the whole. These, to be seen with the naked eye, must be examined shortly after the urine is first voided, as after it becomes turbid they are perceptible only by the microscope, and the urine in this disease becomes turbid in a very short time, because of its highly-concentrated character, for be it remembered that it is always highly concentrated in this affection.

The specific gravity of the urine in this disease, will be above the average; its color will be very high, even to dark-brown, and it will contain curled epithelial scales in large quantities, discoverable by the naked eye before the urine becomes turbid, after which it deposits urate of ammonia, triple phosphates or oxalate of lime, respectively, according to the grade of the disease.

If urate of ammonia is deposited, the addition of liquor-ammonia or liquor-potassa will dissolve it.

If the phosphates, an iridescent film of crystals will form on the surface of the urine. If oxalate of lime, it will dissolve in nitric acid without effervescing.

Whatever grade of inflammation the deposits may be characteristic of, the microscopic appearance will assist in determining.

According to Schonlein, in purely inflammatory diarrhœa, the urine is of a fiery-red color, causing scalding in the urethra, and forms, at the crisis, a crystalline sediment of uric acid. It may be that this scalding of the urine has something to do with the formation of the epithelial scales which are so universally present in these cases.

"In catarrhal diarrhœa, the urine is rather dark, and becomes more so in the evening; at the crisis, a mucus sediment is deposited. In bilious dysentery, the urine is of a dark-red color, tending to a brown; during the crisis it yields a fawn-colored precipitate."

Finally, in typhus dysentery, the urine is dark, turbid, and fetid. During the crisis, it forms no precipitate, but becomes clear and loses its smell."

So you see by authors, that even the character of enteric diseases, the distinguishing marks which make up their difference, whether catarrhal, bilious, or typhoid, etc., may be readily determined by the character of the urine, and hence, by a knowledge of this secretion in disease, the practitioner will be much aided in his diagnosis, and consequently, more successful in the duties of his profession.

Treatment: Of all the remedies in use, the one most universally depended on, and that gives most satisfaction to Eclectics, is the Neutralizing, Cordial, and Sudorific Drops of Dr. Beach; small and oft-repeated doses of podophyllin in combination with three or four-grain doses of Diaphoretic Powder, has been of much service to our patients. Spearmint, flaxseed, or some diuretic tea, should be drank, and sinapisms placed over the region of pain.

INFLAMMATION OF KIDNEYS.

THIS disease, like other inflammatory complaints, is characterized by redness, heat, pain, and swelling of the organs involved. And when we take into consideration the position of the kidneys in the human system, the function they have to perform in preserving the balance of power between the depurating organs, whereby the superfluous matters are removed from the whole of the blood, we can not but regard an inflammation of these organs as fraught with the greatest danger to the animal economy, and liable to produce, suddenly, fatal consequences.

Dr. Beach says: "That diseases are carried off by a copious discharge of urine, every physician knows."

Prof. Eberle says: "At present this excretion is unquestionably too much neglected; by an attention to it, we will often be greatly aided in our judgment of the nature of diseases." It is of primary importance in all diseases in which the urine is concerned, and whatever may be the disease, seldom fails in furnishing us with a clue to the principles upon which it is to be treated."

Dr. Simon says: "From the physical and chemical state of the urine, the attentive, observing physician may obtain a great quantity of information for ascertaining and establishing a diagnosis."

The kidneys are not only charged with the important office of removing from the blood the excess of nitrogen that may be found in the circulation, which can not be got rid of by any other source, but are also charged with the office of removing a great variety of other substances, whether normal or abnormal to the blood, that may enter the circulation or have been retained therein.

When, therefore, her functions are interfered with by inflammatory action, not only the highly poisonous nitrogen will be thrown in upon the blood, which if not speedily removed, will soon produce death, but the various other substances thus retained will also produce serious derangements of the system.

The urine, in this disease, will be scanty and highly colored, at first of a flaming or fiery-red, but soon changes, as the disease progresses, to a dark-brown, is very highly concentrated, and has a strong, pungent, disagreeable odor.

All the fluids of the body become contaminated in this disease, and consequently the urine soon becomes putrefactive and takes on a dark, dirty-red appearance, "becomes semi-solid, when cold," says Simon, "and of a dark color not unlike a mass of black-currant jelly."

In speaking of nephritis arthritica, Schonlein describes the urine as being of a fiery-red color, very acid, and soon after emission depositing glistening crystals of uric acid, in some instances the sediment occupying half the volume of the urine. If the disease terminates in convalescence, at the commencement of recovery the urine is secreted more copiously, and forms a creamy, and often a brown sediment which afterward separates itself into floculent mucus.

The very heavy deposits made in the urine in this disease, are composed principally of nitrate of urea, which is in excess. The average proportion of that constituent in healthy urine is about fourteen and one-half parts in one thousand, and any excess over that proportion is disease, the extent of the excess, as in most cases, measures the extent or severity of the disease also.

When an excess of nitrate of urėa exists, add half the same quantity of nitric acid, place the vessel in a cool place, and crystalization will immediately commence on the edges of the fluid, and soon be thrown down to the bottom of the vessel. They may be readily examined with the microscope, appearing as in Chart, Fig. 1.

Treatment: Mucilaginous and other diuretics, marsh-mallows-root, cleavers, or uva ursi tea, cream-tartar physic, aconite, irisin, leptandrin, etc.

INFLAMMATION OF THE WOMB.

THE womb is a spongy body, composed of arteries, veins, lymphatics, nerves and muscular fibre. Inflammatinn of this organ seldom occurs, excepting after child-birth or delivery, when there are many concomitant circumstances in connection therewith, such as suppression of lochia, secretion of milk, etc., which give additional characteristics to the disease.

Simultaneous with this inflammation and these suppressions, the blood becomes loaded with abnormal substances, the result of these circumstances, for the removal of which, the system is indebted to the urinary secretion.

Dr. Bennet says that "in this disease the urine may be turbid when first passed, and as it cools the turbid matters collect in high-colored flaky clouds, which after remaining a short time in a state of suspension, depose in the bottom of the glass. If not turbid at first, it may become so on cooling. The uriue then is generally of a dark-brown color, and the sediment that forms is also of a very dark-brown, or a dirty-pink hue. These sediments are generally constituted of amorphous urate of ammonia, but may contain crystals of oxalate or phosphate of lime.

The epithelial scales in this disease, as also in that of inflammation of the stomach and bowels, we know is attributed by some to be entirely from vesical irritation of the urinary organs, but it matters not to us from what source, if always present in certain diseases, it serves our purpose as a guide.

If, then, the urine contains not only the epithelial scales, somewhat similar to those in inflammation of the stomach or bowels, but milk-globules as secreted from the blood, and also the dirty-reddish, or pink-hue lochial matters, and in severe cases, where the inflammation has run high, pus-globules, as will sometimes appear, we will have no hesitation in pronouncing the case inflammation of the womb.

The urine is always high colored, or deep and dark-colored, and morbidly loaded with the salts above mentioned, as well as the dirty matters.

If the inflammation progresses, the usual dark, muddy-like appearance of the urine of putrefaction will be presented in connection with the above.

If it is desirable to test the qualities of the deposits, the usual tests for urate of ammonia, oxalate of lime, phosphate of lime, milk, pus, or blood, may be conducted through their several operations with advantage, and the microscope brought to the aid of diagnosis.

Treatment: Sweet spirits of nitre in water, althea-root tea, and tinct. macrotys through the day, with small doses of podophyllin and Dovers powder at night, with rest, quietude, light diet, and wet-bandage around lower bowels.

INFLAMMATION OF BLADDER.

CYSTITIS may arise from an acrid secretion of urine, a predominance of uric acid, from calculi, or whatever interrupts the balance of the circulation.

Tenesmus, pains in the loins, difficult urination, bearing-down efforts, violent pain, nervous irritability, and general febrile symptoms many times supervene. The symptoms of this disease are sometimes so diversified as to lead to unfounded suspicion of many other diseases. We have seen patients treated for dyspepsia, liver-complaint, inflammation of the womb, hysteria, gonorrhœa, gravel, worms, fits, and even fever, who were only laboring under cystitis.

One case, a female aged twenty-five years, who was not out of bed for nine months, and who was treated all that time for inflammation of the womb, by a celebrated physician, who had tortured her with the actual cautery over the bowels, so much that he could no longer heal the wounds his "hot-iron" had made. He seems, however, to have a kind of mania for the actual cautery, as we are told he has his iron in the fire nearly all the time, and cauterizes for nearly every disease.

Another case, a little girl aged ten years, who was so violently afflicted as to be sometimes convulsed. She was treated successively by three physicians for fever, for worms, and for epilepsy, neither of them having examined the urine, nor ever once suspected the true cause of the disorder.

A number of other cases of this disease have came under our notice, which we have suddenly relieved, because of the true nature of the case being made known to us through the urinary deposits.

Although vesical mucus may be nearly always present in healthy urine, in these cases it is exhibited in large quantities, in an irregular, white gelatinous mass, which adheres to the bottom of the glass, or floats through the urine in an aggregated form which can not be dispersed by agitation, nor mixed with the urine.

This characteristic is alone sufficient to determine it from pus, which it resembles.

To test it, add a little acetic acid, which will coagulate it into a thin membrane. Sulphuric acid will not dissolve it, and ferro-cyanide of potassium throws down a precipitate. For microscopic appearance, see Chart, Fig. 11.

Treatment: Diuretic emulsions and mucilaginous drinks, combined with syrup of poppy, and other sedatives.

INFLAMMATORY RHEUMATISM.

THIS, the last on the list of inflammatory diseases which we shall take upon ourselves to diagnose by the urinary secretion, is one in which the theory, as well as the practicability of our peculiar system, is made manifest to every observing practitioner as being substantiated by the highest authority.

"After the most mature consideration of this subject," says Dr. Robertson, "based not only upon what has been written by the most experienced authors, but also upon the results of minute observatins in practice, we long since came to the conclusion that genuine uncomplicated rheumatism is always connected with an excess or redundancy of the fibrinous constituents of the blood.

"The state of the blood, too, when drawn, very generally shows that there is a redundancy of fibrin."

Dr. Buckler says: "In acute rheumatism the disease is seated in the blood."

The blood in acute rheumatism, accompanied by fever, always exhibited the character of inflammation. In the period of acute pain and fever, the fibrin always existed in much larger proportions than in normal blood, and if, by repeated bleeding, the fibrin was reduced in quantity, upon the subsidence of pain, in those cases in which the pain and fever returned, an increase of fibrin was again observed, according to the experiments of Andral and Gavaret.

The acuteness of pain seems to have as great an influence on the increase of fibrin as the stage or duration of the disease, the blood containing as large an increase of fibrin at the commencement of a fever attack as at a later period in a milder case.

Simon says: "We have already seen that the blood in rheumatism perfectly corresponds with the blood in true inflammations; hence we are led to infer that the urine will also present the inflammatory type—an inference confirmed by experiment."

It will exhibit in the acute stage those characters of inflammation which have been so often presented, in a very high degree, the extent depending upon the violence of the attack.

As the rheumatism becomes chronic in its character, and without fever or inflammation, as Eisenman remarks, the properties of the urine may undergo a change, and instead of being acid it may assume an alkaline reaction, and give off a fetid ammoniacal odor.

Urine in rheumatism often throws down sediments, even at the hight of the disease, when the deposits formed may be regarded as significant of a true crisis, if the supernatant urine is perfectly clear.

The disease, then, being seated in the blood, and having found *one* constituent always to be in excess in that fluid, under these circumstances we will expect, according to the physiological function of the kidneys, to find an excess of that substance thrown off by the urinary secretion, when they perform their functions aright; and such we find to be the case.

But there are other signs in this important condition, which give farther clue to the nature of the case.

"As surmised by physicians in all ages," says Dr. Lewis, "it is apprehended that by acting on the blood we can really reach the *materis morbi* in rheumatic affections; the modern opinion of an excess of uric acid in the blood being, if not the proximate cause, at least mainly instrumental in constituting the disease, is confirmed by the effect of remedies that relieve it."

Colchicum appears to be the great remedy, and all agree that the uric acid deposits in the urine are always more than doubled under its use.

So fully satisfied have physicians become of the excess of urea and uric acid in the blood of rheumatic patients, that some have christened the peculiarity with the technical term of *diathesis*, uric acid *diathesis*, or disposed to excess of urea in the blood, and deposits of uric acid or urea in the urine; hence persons disposed to acute rheumatism are those having that excess.

In this disease, therefore we will always find the general characteristics of inflammation presented in the quality of the urine, together with an excess greater or less of these deposits. The urine will be very highly colored, tinged with reddish floculi, and loaded, as it were, with a thick, dense cloud, visibly suspended in it, and on cooling will become turbid, finally, will throw down such large quantities of uric acid, as said by Dr. Buckler, "resembles the washings of a wine-cask or beer-barrel." This very great excess, however, generally takes place only on the crisis forming, or at the period of convalescence.

The urine in this disease is always acid, hence it reddens litmus-paper. "It presents every shade of tint," says Bird, "from the palest fawn-color to the deepest amber or orange-red," hence the name of red sand, is applied to these deposits.

The specific gravity of the urine in this disease, is always above the average, and, as in a generality of cases of urine, the deeper the color the greater the density, and vice versa, the lighter the color the lighter the specific gravity. The practiced eye will soon detect, with great precision, the probable weight of the urine, without the aid of a gravimeter.

For microscopic examination, if a great number of crystals are desirable, by the addition of a little hydrochloric acid, a great abundance will be deposited.

This examination may be conducted in the manner as recommended by Bird, or as given in treating inflammatory fever, to which this is very nearly allied, which see.

This closes our examinations of inflammatory diseases, any one of which here mentioned, may be easily determined by inspection and analysis of the urinary secretion by the tests, reagents, and microscopic appearances.

Every one should be well acquainted with the appearance and composition of urine in a normal state, before entering upon the practical application of its principles to medicine. We should be prepared to recognize it in every condition, and be acquainted with each of the constituent principles in all the several changes. For, "more than all other signs," says Simon, "the correct examination of the urinary sediments, is of importance to the physician. Healthy urine forms only, after long standing, a light, sinking cloud of vesical mucus; every other separation in the urine is of a pathological nature."

Treatment: Diuretic medicines, the alkaline salts, or whatever defibrinates the blood, eliminates or dissolves and discharges the uric acid, and removes the peculiar diathesis.

Nitrate of potassa has been highly extolled in this disease; but iodide of potassa, in connection with nervine tonics, seems to be more effectual, however. Colchicum is undoubtedly a good remedy, in many cases, but will require nervine tonics in connection. Macrotin, however, seems to possess the properties more in combination, and to be more effectual in removing this complaint than any other single remedy; or the tincture of macrotys racemosa, which we prefer to the concentrated preparation.

Whatever the remedy, it is required to begin in full doses, even to the production of their own peculiar characteristic symptoms or constitutional effects.

The treatment of acute rheumatism and all inflammatory diseases, by nitrate of potassa, would seem, however, to be of considerable pathological interest, since it is asserted, upon good authority that it is a powerful solvent of fibrin.

"We know," says Simon, "that nitrate of potassa is a powerful solvent of fibrin," and we are thus enabled to interpret the efficiency of this remedy, in accordance with our knowledge of these diseases.

FEVERS IN GENERAL.

We are fully satisfied that the existing causes of all diseases are first manifested in the circulating fluid, the blood; and especially is the proof abundant in the case of the different fevers, so called, whether the cause be in the change of its proportionate constituent principles, or in the deleterious, abnormal or morbific agents introduced therein.

"The seat of fever," says Beach, "is in the blood-vessels, or vascular system, hence it is evident that the exciting cause must be in the blood." It has been shown by many pathologists that the blood in different forms of fever is in a morbid or vitiated state. It has been subjected to chemical analyses in nearly every disease, and all tend to show that previous to the other signs of attack even, the blood is materially altered in its properties, and that its constituent principles undergo progressive changes as the disease progresses. Hence, Dr. Stevens considers this diseased state of the blood as the first link in the chain of morbid phenomena which constitutes fever. He believes that the ærial poisons from which all pestilential diseases arise, are carried with the atmospheric air into the circulation, mix directly with the blood in the pulmonary system; and that this poisoned or diseased state of the whole circulatory current is the cause of the subsequent morbification in the solids.

"The blood is not only materially changed in fever," says Beach, "but the diseased state of the blood precedes the attack, and the changes take place in a determinate order."

Laurentius Bedine thus defines fever: "Fever is a faulty state of the blood in either motion, quantity or quality, or in all of these respects."

Now, the most frequent alteration in the character of the blood in all idiopathic fevers, as they are called, is known to consist in the diminution of the natural proportion of its fibrin, "and that the degree of this diminution is very uniformly proportioned to the severity of the disease," says one author.

This, it is observed, is the reverse of that which is always found in inflammatory fever, as may be seen by referring to that subject, as treated in preceding pages; and is an important fact which is necessary to be borne in mind in all the examinations. In proof of this, read the following:

"It is obvious," says Carpenter, "that increase of fibrin in the blood does not exist as a result of these fevers. In typhoid fever the decrease in the proportion is much more decidedly marked, and as a favorable change occurs in the disease, it again approaches the normal standard."

Dr. Andral has observed a marked diminution of fibrin in the blood in all idiopathic fevers.

Without going into a history of all the phenomena of the different kinds of fever, it will be sufficient for our purpose in this place to state that, in all idiopathic fevers, there is not only a decreased state of the blood, but that there is a derangement of the secretions also, as is indicated by the dry tongue, increased thirst, hot skin, and scanty and high-colored urine, which is altered in appearance and constituents. "It is the case," says Beach, "that fevers and other inflammatory diseases are brought to a favorable termination by a spontaneous discharge of urine."

In view of this universally altered state of the blood, and its difference from other diseases, and also the corresponding alteration in the constituents of the urine at the same time, we have only to learn in what that alteration in the urine consists, to determine from that secretion, the character of the fever.

The urine then, in idiopathic fevers, is generally "scanty and high-colored," but we must remember that it is unlike that of the inflammatory, in being destitute of the "reddish-tinged floculi," which added brightness to the color in the latter. There will not be even a normal quantity of that substance found, but rather a proportionate deficiency, which deficiency is generally made up in the abnormal quantity of other matters.

The high color of the urine is partly occasioned by the degree of concentration which that fluid takes on, because of its scantiness, together with the peculiar substances transmitted to it from the circulating fluid. It is these substances that give to each specimen a distinct peculiarity of its own, according to the kind of fever, thus: The chemical pathology of bilious or remittent fever, reveals the fact, that this disease spends its action more upon the liver than upon any other organ; while that of intermittent fever is spent more upon the spleen, as indicated by a tenderness over that region, and an enlargement of that organ in protracted cases; and that of typhoid fever more upon Peyer's glands, and so on, as is well known to all medical men.

Now, in each of these fevers, while we will find the urine always highly, or we might say, deeply colored, measurably, because of its great concentration, the color somewhat resembling "strong ley," and deficient in the metamorphosed fibrinous matter, and also certain other peculiar substances, according to the organ or part predominantly affected, as the liver, splean, Peyer's glands, etc., in accordance with these characteristics will we find the type of the fever.

These several peculiarities will be noted under the respective head to which they belong. We will close this article on fevers in general, with a few quotations found in Beach's American Practice, Vol. II.

1st. A deranged state of the secretions invariably precedes the occurrence of the slightest degree of excitement. This we think will hardly be disputed by any one.

2d. The pre-existence of such a condition being admitted, its adequacy to promote the excitement can not, we suspect, be questioned.

3d. In proportion as the secretions become re-established, the excitement invariably diminishes.

4th. The excretions, upon the decline of the fever, invariably contain a much larger quantity of salts and other matter than they do in their ordinary state. We need only refer to the urine, which is so loaded with these substances, that a copious deposit takes place almost immediately after it has been voided. This affords an additional proof that there does actually take place, during the progress of fever, an accumulation in the blood of matter, which in a state of health, is regularly excreted by the kidneys.

INTERMITTENT FEVER.

THE cause of intermittent fever is now pretty generally conceded to arise from what is termed *malaria*, a certain poisonous gas, which mixes with the blood, and becomes so obnoxious to it that a new action is set up in the system to expel it from the circulation.

We need not mention all the strange phenomena that generally characterize this fever, but only point out some of the never-failing concomitants which distinguish it from other diseases, the principal among these are: regular and distinct periodical return of chills, and the universally deranged condition of the spleen.

By whatever influence the blood may have become contaminated in this disease, whether by malaria or other poison, or by a deficiency alone of the fibrin of the blood, it is certain that a deranged condition of the spleen is universally present.

Maillot thinks that in most cases there existed some lesion of the spleen before the accession of the fever. Dr. de la Bac refers the element of periodicity to the introduction into the blood of an acid (probably malaria), and Dr. M. Roche, to the contamination of the blood by malarial poison.

Now, it appears to us that the deranged condition of the spleen alone (perhaps because of the malarial poison), gives rise to the periodicity, as well as the other equally mysterious phenomena belonging to this disease.

The spleen is a spongy viscus, of considerable magnitude, and is connected by ligaments, vessels, etc., with the stomach, kidneys, omentum, diaphragm, peritoneum, pancreas, colon, etc., and yet, "its use," says Hooper, "has not hitherto been determined, at all events, is very little known." But of its derangement in this affection, as well as its altered structure in protracted cases, no one can for a moment doubt.

Of all the organs of the body," (says Prof. Paine, in speaking of intermittent fever,) "the spleen is the most uniformly affected. It is not only changed in structure, but becomes very much enlarged, and in all fatal cases proves to be the principal organ on which the disease has spent its force."

"In the worst forms of intermittent fever," says Beach, "the effect is aggravated by the great intensity of the malarial poison; the blood not only stagnates but accumulates in an altered state, prone to decomposition, and in these cases the spleen is represented by writers as resembling a bag of tar, the blood being as black as pitch."

"There seems to be some noxious influence produced by the spleen, in a diseased state, which impairs the making of blood; the large diseased spleens are like bags of spoiled blood, the return of which into the circulation injures and reduces the other blood, and keeps up a continued disease."

"The spleen, in periodical fever," says Bartlett, "is almost always enlarged, softened, and of a very dark or bluish-black color. This lesion is so constant, and has been so long familiar to all observers of this class of diseases, that it is hardly necessary to multiply cases, or to quote authorities."

"We have often seen these subjects," says Bailey, "arriving at the hospital in Rome, with abdomen as hard as a stone, the spleen occupying the whole anterior part of the cavity."

Enough, we think, has now been given on this subject to satisfy anyone that in intermittent or periodical fevers, the blood is primarily affected with a tendency toward spending its action more upon the spleen than upon any other organ, and that it is to this peculiar change in the blood, producing this effect upon the spleen, that gives the otherwise unaccountable periodicity to this fever.

Now what are the facts in regard to the state of the urine, in this disease, and do they bear any relation to this view of the case?

Prof. Eberle says: "In the cold stage of this fever, the urine is clear, colorless, without sediment and often copious; in the hot stage, scanty and deep-colored, without sediment. In the sweating stage, the urine, though still very high-colored, deposits a lateritious or pale-red sediment."

It will readily be observed here that, in each of these stages there is found to be a change in the character of the urine, corresponding with the change in, or stage of the disease.

Dr. Bartlett says: "In the cold stage the urine is generally abundant and limpid; in the hot stage it is scanty and high colored; and in the sweating stage, the urine deposits a reddish sediment."

Dr. Beach makes very similar statements to those of Eberle and Bartlett, in regard to the urine in this disease; as also does Prof. Paine, who adds that "this fever is generally preceded by alternate changes in the urine, which characterize the type of the approaching disease, by occurring at the same regular intervals and lasting the same length of time."

Dr. Lenac lays still greater stress on the appearance of the urine, as characteristic of periodical fevers. "Masked intermittents," he says, "may be no less certainly detected, as was formerly observed by the color of the urine. In this disease the urine is very often lateritious during the remission, *which is a sign almost infallible*, that it belongs to this family."

According to Dr. Currie, "the urine during the cold stage is pale, copious, and crude, but as soon as the hot stage is established it becomes high-colored, while in the remission it is thick and cloudy; and sometimes deposits a brick-colored or brownish mucous sediment."

Dr. Boling says that the secretion of urine is scant, highly colored and muddy, during the exacerbation, from the coloring matter floating suspended in it; and sometimes late in the disease is of a deep reddish-brown, possessing apparently a degree of consistency greater than natural, etc.

What remarkable coincidence of opinions among learned authors, without having attempted any philosophical, or pathological explanation of the causes, or ever once attributing the "universally the same" condition of the urine under like circumstances in this disease, to the uniformly diseased state of the blood, and to the uniformly alike-affected condition of the spleen in the same disease!

It may appear that enough has already been said to enable anyone to detect a case of intermittent fever, by the characteristic appearance here given of the urine alone; and we shall only mention in conclusion, the facts that the spleen in this disease, according to Paine, always yields a peculiar elastic resonance, and that according to Bird, the presence of an excess of purpurine in the urine is almost invariably connected with some functional or organic mischief of the liver or spleen, and that the pink deposits which are almost constantly present in the urine (in these cases), are always to be found in the hypertrophy of the spleen following ague.

There is a deficiency of fibrin in the blood, and consesequently a deficiency of this element in the urine. The urine is always alkaline to test-paper, and contains urate of ammonia in excess. (See Chart, Fig. 5).

"The sediment formed in the urine in intermittent fevers, is always composed of uric acid and urate of ammonia, in most cases combined with red coloring-matter, (uroerythrin.) (Simon).

The high color of the urine is generally owing to excess of purpurine—if so, alcohol will take up the purpurine, leaving the urine more clear. Urate of ammonia disappears with slight heat and reappears on cooling; and liquor-ammonia or potassa will dissolve it when in deposit.

A microscopic examination of urate of ammonia may be made by placing a drop of the turbid urine between two plates of glass; an amorphous powder will be visible under the microscope, which will be readily recognized by comparison with the figure representing it in the Chart, or with a "prepared specimen."

Treatment; Remove the congested state of the spleen, and the circulating fluid thus: 1st, Give an emetic of Euphorbin and Lobelin, then small doses of podophyllin and eupurpurin. 2d, Relieve the *periodicity* with cypripedium, salicin, cinchonin and acidulated drinks. 3d, Tone up the system with lemon-peel, iron, hydrastis and wine—to guard against, or or prevent a relapse.

If the chronic enlargement of the spleen has supervened, an irritating plaster over the region will be of service.

REMITTENT FEVER.

We have dwelt longer on intermittent fever, because of one of its main peculiarities (periodicity), being, in many cases, a partial characteristic in this and other fevers. In fact, in whatever disease, if the spleen in any way becomes involved, periodicity will be one of the concomitant symptoms. Hence we will not unfrequently find it participating in a slight degree in this, because of the spleen being partially affected also.

Of all the organs of the body, however, the liver is the one universally affected in remittent fever. "There is good reason for believing," says Bartlett, "that the lesions of this organ constitute the anatomical characteristic of this disease."

When we consider the position of the liver, its office as a depurating organ, its importance in manufacturing bile, as it were, and its general relation to the process of purifying the blood; and that in remittent fever the liver, more than any other organ, is the one most uniformly and constantly affected, it is certainly to the peculiar phenomena produced by such affection, that we are to look for a solution of the nature of the disease.

The natural color of the liver in health, is a deep red, or reddish-brown. In remittent fever Dr. Robertson made post-mortem examinations in a great many cases, and found the color to be nearly the same in every case, but very different from the natural.

In most of the cases he described the liver as being the color of bronze, or a mixture of bronze and olive.

"The most correct idea of the color before us," says Bartlett, "perhaps would be conveyed by stating its predominant character, the same in every case, to be a mixture of gray and olive, the natural reddish-brown being entirely extinct, or only faintly to be traced."

Dr. Gerhard's cases in the Blockley Hospital presented the same appearances. In all of Dr. Sweet's cases in the New York Hospital, the peculiar change in the color of the liver, described by Dr. Stewardson, were present. Also, in the cases summed up by Dr. Stille; "in all, the color of the liver was either bronzed or like that of slate."

Dr. Stewardson very naturally suggests that this alteration of the liver may be found to constitute the essential anatomical characteristic of marsh-fever, as the lesions of Peyer's glands and the lymphatic ganglia constitute that of typhoid fever.

In this fever, the blood is changed in quality, and rendered both chemically and mechanically unfit for the purposes destined by nature, and is different in character from that of any other fever; the peculiar character of the lesions being different, as the foregoing shows. It is sometimes termed bilious fever, because alone of the bile which enters the circulation of the blood and thereby changes its constituents, pervades the tissues and gives a yellow-color to the skin, is secreted from the blood by the kidneys, and alters the chemical qualities and appearance of the urine.

The urine in bilious fever is usually impregnated with bile-pigment; it is of a more or less brownish color, and when a thin layer is seen it appears of a citron-yellow tint; it differs, however, with the degree of vascular excitement. The presence of bile-pigment may always be recognized by the change of color which succeeds the addition of nitric acid.

"The urine is thick and of a dark-yellow color," says Paine, "and in the latter stages scanty and muddy."

"The eyes and skin assume a yellow-tinge," says Beach, "and the urine is scanty and of a yellow-color, and mixed with bile."

In bilious fever then, when the leisons of the liver universally present the appearances represented by eminent authors here mentioned, and the blood is so uniformly altered in constituents, there being a deficiency of the fibrinous

substance, and an abnormal proportion of the biliary matters, the urine being secreted from this state of blood, we could not philosophically or in any other way, expect to find a normal state of that fluid during such a state of the human system.

But, on the contrary, we find the urine participating in the same characteristics of the disease, in so prominent a manner as to identify it fully in every case.

The urine not only takes on the peculiar brownish or bronzed appearance observed in no other disease, but it partakes largely of the contaminating matters with which the blood has become surcharged, because of the unhealthy bile entering the circulation, and hence will exhibit the greenish or bilious tinge also.

"Bile often tints the urine of a deep-brown color, also," says Bird, "and may lead to an unfounded suspicion of blood."

But no one at all skilled in the ocular inspection of urine in disease, will ever be deceived in this for a moment; yet, if there should be a doubt at any time, the very convenient nitric-acid test will soon decide.

It would be well for every learner to obtain a specimen of urine from a well-marked case of bilious fever, and in fact, every other disease, seal them up and preserve them for frequent examination and comparison, to familiarize themselves with the appearances.

The deep, heavy-brownish or bronze-tinged urine, which is hard to describe, because of its peculiar mixture of coloring matters, may be very readily recognized, however, by a comparison with other urine, or with the urine of health even.

Its chemical constituents, however, will more readily detect or determine the extent of the disease by the quantity of abnormal, or excess or deficiency of any constituent.

It generally deposits purpurate of ammonia in abundance, in connection with a peculiar substance, the disintegrated particles of the liver itself, they being of the grayishor bronze-color, before mentioned.

When the spleen is at all implicated in this fever, there will be traces of urate of ammonia in the urine, as in intermittents, and the fever will be somewhat marked by the usual symptom of periodicity. These substances may be readily recognized by their tests, and the microscope; if bile, test with nitric acid.

The urine will be alkaline to test-paper; specific gravity will be high; will be reddish-brown, or mixed with bronze or gray, and will deposit purpurate of ammonia, with perhaps urate of ammonia.

Treatment: First, an emetic of Lobelin and Euphorbin, then moderate doses of leptandrin, oft-repeated through the day; and podophyllin and diaphoretics at night, with dandelion-tea throughout the disease.

TYPHOID FEVER.

Of all the fevers which prevail in this and the northern portion of our country, there is none to be dreaded more than typhoid fever.

Its peculiar lesions have given rise to various names for this formidable complaint, such as entero-mesenteric, abdominal-typhus, intestinal-ulcerating-typhus, typhus-gangliaris, etc.

As it is our object only to give the pathological condition of the blood so far as it relates to the consequent changes produced in the urine, and the lesions or other peculiarities that may have a bearing upon that fluid, we will proceed to the special case.

"The most frequent alteration in the character of the blood in this fever," says Bartlett, "consists in the diminution of the natural proportion of its fibrin."

Andral and Gavarret found that in typhoid fever the proportion of fibrin in the blood was never increased above its natural standard, but on the contrary, that in many cases this proportion was very much diminished; and, furthermore, that the degree of this diminution was very uniformly proportionate to the severity of the disease. Here, then, is the measure of the severity or extent of the disease itself, in whatever measure of deficiency of fibrin from the natural standard in health.

The blood in typhous abdominalis exhibits a diminution of fibrin from the commencement, and that element never rises above the normal standard, in true typhoid. This decrease of fibrin is almost always connected with an increase of the blood corpuscles and solid constituents, also, until the disease approaches the stage of collapse. As this stage is approached, the fibrin still decreases, but a remarkable change comes over the blood corpuscles and the solid constituents (in consequence of the liquor-sanguinis being too watery and deficient in salts), and these decrease also. As soon, however, as any symptoms of convalescence appear, the fibrin begins to increase, and this continues to be the case during the progress of convalescence, during which progress the corpuscles simultaneously and correspondingly decrease. Now, these changes in the blood will give corresponding changes in the urine, and M. Primavera, who has been observing the constituents of urine, makes the following statement in reference to the urine in typhoid fever:

1st. The complete absence of the chlorides from the urine is a pathognomonic diagnostic sign of typhoid fever. This valuable sign will serve to distinguish this fever from a simple and benignant fever, continuous or intermittent, in which the urine always contains an appreciable quantity of salts of this nature.

2d. Urine passed during the ascending period, or even during the whole course of typhoid fever, when this has a fatal issue, shows not only an entire absence of the chlorides, but even a very considerable diminution of the phosphates and urates.

3d. The first step toward convalescence is indicated better than by any other sign, by a very rapid and sensible increase of the phosphates.

4th. The second phase of amelioration is shown by an analogous increase of the urates.

5th. Finally, the re-appearance of the chlorides in the urine, however tardy, definitely indicates the recovery of the patient.

The peculiar lesions which characterize this, is different from every other also. "In all cases of typhoid fever," says Bartlett, "there is lesion of the small intestines. This lesion is peculiar. It is found in no other disease. It is generally extensive."

He then goes on to state that constituting, as this lesion does, the characteristic, and of course, the most interesting and important pathological element of typhoid fever, he will describe it with all possible accuracy and completeness, etc.

The invariable and characteristic lesion found in the small intestines, consists in alteratiens of the elliptical plates, or of what are called "Peyer's glands." These plates or glands are more or less extensively the seat of ulcerations, sometimes amounting to a perforation of the intestines, through which their contents may be discharged into the peritoneum.

Peyer's glands are sometimes so morbidly affected, or changed in this fever, that there is deposited in the cellular tissue a substance of a yellowish-color, that has been denominated typhous matter.

Dr. Bartlett says that the conclusion to which he is irresistibly led, is that the connection between the diagonistic symptomatology of typhoid fever, and the entero-mesenteric is, we will not say absolute and invariable, but as nearly so as the connection between the diagonistic symptoms and the characteristics lesions of any disease whatever, in the nosology.

What is the nature of this alteration of "Peyer's glands?" asks the same author, and concludes by attributing the lesion to inflammation, not of a common, but of a specific character, having something peculiar in its nature, in regard to the condition of the blood; and that this fever consists essentially in this lesion, and refers the symptoms to the lesion as their cause, the blood being affected thereby.

In all cases wherein the typhoid state or the typhoid element in pathology has been present, important and peculiar changes have taken place in the blood; "therefore, there is good reason," says an author, "to think that the changes thus produced upon the blood play a very important part in the pathology of the disease."

The peculiar alteration in the blood, whether produced by the action of malarial poison directly upon the blood, or primarily, upon Peyer's glands, thence upon the blood, matters not, for our purpose, so long as the effects upon these glands and the blood are so simultaneously and uniformly present; for it is in the urine secreted from this diseased condition of the blood, that we are to rely principally for the diagnosticating substances passed therefrom.

"The kidneys do not only secrete" says Lehman, "certain constituent atoms of organs, which have become unfitted for the purposes of life, but also the excess of nutritive matters," etc.

"M. M. Solon," says Dr. Wood, "has paid particular attention to the state of the urine in this disease. He has found it to be more scanty, higher colored, and denser than in health, equally acid, if not more so; much more abundant in urea, and occasionally albuminous, especially in severe cases, in which this character of the secretion must be considered an unfavorable sign."

It is well known that retention of the urine is very frequent in this form of fever, and is one of the particular marks which distinguish it from other forms of continued fever.

Dr. Nathan Smith says: "In the commencement of this fever the urine is not high-colored, and is considerably copious, being often above the natural quantity, and deposits no sediment. In voiding it into a glass it often foams like new beer. As the disease advances, the urine becomes more highly colored; and as it begins to decline, lets fall an abundant sediment."

Drs. Dobler and Skoda inform us that whenever the disease is at all severe, the urine deposits no sediment, unless it be a slight mucous cloud, but on the subsidence of the fever there is often a dirty-grayish deposition.

It appears from the observations of Schonlein and Simon, that in typhoid fever the urine is at first very acid, subsequently, neutral, and even alkaline, and again acid at the commencement of convalescence.

Carbon, azote, hydrogen and acids, all seem to be retained in the blood in this disease, because perhaps of the partial retention, or scanty secretion of urine; and the perspiration, urine, and excrements, are all generally acid.

The urine will be acid, especially in the first stage; it will be loaded with the "bead" or bubbles of carbonic acid, will be tolerably clear until the second stage or greater violence of the disease, when it will be found more highly or darker colored, and will emit a very strong odor.

In the latter stage when ulceration of the glands of the intestines has supervened, the urine will become cloudy and thick with a mucous-like substance, which, on repose, deposits a dark muddy-like matter, indicative of serious disorganization of these parts, and dangerous dissolution of the blood.

So plainly is this condition of the system manifested in this appearance of the urine, that no prognostic sign of approaching death can be more safely relied upon. A prognosis of the approaching end of life can be definitely made out from the peculiar dark or blackish muddy-like appearance of the urine, sometimes several days or a week even before the final cessation of life—as we have often been called upon to do.

This peculiar characteristic of the urine, although not easily described, yet when once observed scrutinizingly, will forever after be remembered.

Treatment: Avoid drastic purgatives, restore the suppressed secretions, especially the urinary, and keep up the tone and strength of the system. Give leptandrin, terebinthinate diuretics, and Peruvian-bark tonic, with alkaline sponge-baths, and soda drinks.

YELLOW FEVER.

In this disease we have had no practical experience whatever, and therefore can not be expected to give much valuable information concerning the state of the urine therein.

From the history of this much-dreaded disease in warmer climates, it seems, however, that some deducible facts may not be out of place.

The blood in this disease is described as watery, very poor in fibrin, and of a dark color. If any clot be formed, it is diffluent, and very soft. The serum is of a deep yellow or brown-red color, partly from the coloring matter of bile and partly from dissolved hemato-globulin. It possesses a very peculiar smell, and Schonlein and Chomel both speak of a peculiar gas developed in the veins, which escapes with the blood, in post-mortem examinations; Ancel remarks that in the first stage of this fever the blood is of a brighter red, contains more salts, and is hotter than in a state of health. As the disease progresses it loses its saline and animal principles and becomes black and thin.

Balard is of the opinion that the lymphatic system is first disordered, and that inflammation, degeneration, and suppuration of the lymphatic ganglia and vessels follow. When the disease is fully established the arterial and venous blood have both the same dark color, and appear in a peculiar state of solution.

All those pathologists who have written on the subject of this fever, agree that the most important element in its pathology is to be found in the peculiar alteration of the blood, together with a peculiarly striking and uniform change in the color of the liver, wholly different from that of the same organ in any other disease, and also that there is nearly always an altered state of the mucous membrane of the stomach.

The blood was found by Louis and Trousseau to be generally either liquid only, or liquid and clotted; the clots being black, or yellow and fibrinous.

Dr. Nott, of Mobile, found the blood dark and fluid in every case where the bodies were opened. He says it always shows a great deficiency in fibrin, and in some cases it did not coagulate at all.

The uniform change in the color of the liver, is described by some writers as being of a straw color; by others, of a yellowish gum color, a mustard color, a buff color, a pale-yellow color, etc., and generally hard and destitute of blood; and the stomach generally contains the matters of *black vomit*, which are considered to be vitiated blood mixed with the altered secretions of the stomach derived from the mucous membrane.

Now, in this state of the stomach and liver, together with the peculiarly altered condition of the blood, in which so invariably and essentially consists the characteristics of this fever, is it not highly probable that the urine which is secreted from this blood will partake of the qualities of that fluid, and the organs thus affected? And, is it not passing strange that so little attention has been paid to that secretion in this disease, unless, like in cholera, there is generally an entire suppression of urine, which perhaps is the case.

Be that as it may, however, we think if a sufficient quantity of that fluid could be obtained for examination, it would exhibit all the marked indications of the disease, and would prove a valuable prognosticating symptom, at least, and might lead to profitable information respecting the treatment of this formidable complaint.

In Bryant's account of the yellow-fever in Norfolk and Portsmouth, Va., 1855, (whither Dr. Finch, who was practicing with me at the time, proceeded, at the instance of the "Board of Trade of Pittsburgh," to assist in its treatment), it is reported that there was one symptom from which death could be invariably prognosticated, to-wit: suppression of the urine. It was a certain indication of the general dissolution or breaking-up of the animal functions. Perhaps not one patient in whom it occurred survived. This suppression often occurred in the first stage of the disease, and yet it still remained the most certain symptom of dissolution.

This same tendency of the suppression of the urine was observed by us in cholera, during that epidemic here in 1850. And we remember a number of cases of that disease, in which, on the re-establishment of that secretion, and on a recurrence of a plentiful flow of urine, convalescence ensued.

Might not the efforts of physicians have been directed toward the re-establishment of that secretion, with advantage in yellow-fever, as we are persuaded it was to many in cholera, during the epidemic in Pittsburgh?

PUERPERAL FEVER.

This disease is entirely different in its pathological condition from any of the before-mentioned idiopathic fevers, and in fact, the condition of the blood in some respects, is the reverse of all, excepting the inflammatory diseases. Indeed it seems more properly to belong to the inflammatory class of diseases, than to be regarded among the fevers.

The blood in this disease has been analyzed by Heller: it was found to be of a very dark-brown color, but coagulated in the ordinary manner; the serum was turbid, but after standing for some time became clear. The clot was dark, of a loose consistence, and covered with a strong, buffy-coat, over which there was a delicate membrane that presented, under the microscope, a finely granular appearance, and fat vesicles.

Dr. Scanzoni says: "The writings of Helm, Kiwish, Rokitansky and Engel, show that the blood in puerperal fever always contains an excess of fibrin, and that this excess of fibrin not only constitutes the primary cause of the disease, but that it is, in short, in itself, the essence of genuine puerperal fever."

In the vital fluid, the blood, is found the first manifestation of this, and every other disease. "From the vitiated condition of the blood," says Ramsbotham, "this malady derives its extraordinary malignancy."

Rigley considers that the affection commences in the blood. And the German schools consider this disease to depend upon a metastasis of the blood destined to form the secretion of the milk, from the breasts to the peritoneum. It is upon the peritoneum that this disease spends its action, like the spleen in intermittent, the liver in bilious, and Peyer's glands in typhoid fever.

This disease runs its course in a very short time, being of a very malignant character, the blood very soon passing into a highly putrefactive state—probably from the absorption of the secretions of milk, urine, and lochia, which early become suppressed.

The morbid condition of the blood, however, is somewhat allied to that of the blood in erysipelas, and the urinary secretion in these diseases, and seems to be somewhat allied in constituents also. And, in fact, many authors assert that contact with a puerperal patient may superinduce an attack of erysipelas in persons predisposed to that affection.

Puerperal fever seems to be not only communicable from lying-in-women to patients of the same class, but from the children of puerperal patients being so frequently attached with erysipelas, and from which they generally die, it is believed that erysipelas is superinduced thereby.

The urine in this disease is different from that of any other fever. There is an increased proportion of fibrinous substances, with, sometimes, the reddish-tinged floculi noticed in inflammation; this latter is sometimes abundant and gives to the urine a bright and very highly-colored appearance.

Scherer analyzed the urine in a number of cases of *febris puerperalis*, and found it usually of a fiery-red color, sometimes neutral, often alkaline, and depositing a mixed sediment of pus, mucus, and urate of ammonia. In one case, however, it resembled buttermilk, and was loaded with urate of ammonia.

Bouchardet has published an analysis of milky urine, passed by a woman with this disease. It contained no traces of sugar, milk or casein, the appearance being due to a large amount of urate of ammonia.

Ramsbotham says the urine is generally very defective, high-colored, turbid on cooling, and voided with difficulty or pain.

In the cases we have had an opportunity of examining, we have always found it a very highly and bright-red color in the first stage of the disease; and as the disease progresses the urine becomes more thick and turbid, finally muddy-brown, and lastly, dark and muddy or coffee-ground color.

When this latter appearance is observable, all hope of recovery may be abandoned, as this is a mark, unmistakable, of the putrefactive state of the blood.

The urine is generally of a very offensive odor, from the very commencement of the disease, becoming more and more so, during the increasing violence of the same.

The urine in this disease is of high specific gravity; is acid to test-paper, or neutral; deposits uric acid crystals, and urate of ammonia, and sometimes pus-globules are observable by the microscope.

Treatment: Re-establish the secretions of milk, urine and lochia, and induce perspiration; equalize the circulation and remove the congestion of the peritoneum. Give diluent and diuretic drinks, alkaline diuretic salts, neutralizing cordial, chlorate of potassa, tonic bitters, etc.

SCARLET FEVER.

THIS disease, like the preceding one, in our opinion, belongs to the inflammatory class, or perhaps still more properly to the eruptive; yet its description may not be far out of place here.

It is considered to be propagated by a specific, unknown poison, the primary location of which is in the blood, and for the elimination of which poison, nature seems to make use of the skin, as is evident in the peculiar eruption; and of the kidneys, as is evident in the peculiar character of the urine. And, according to the experience of Dr. Bird, the better the development of the rash, the less the embarrassment of the kidneys, and vice versa. The kidneys, he says, are the organs by which the matters in solution in the blood are usually excreted, and in these cases, if there has been a deficient determination to the skin, the kidneys are made to do a supplementary duty.

The blood from the very beginning appears to be loaded, as it were, with a morbid poison, and as the disease advances, if this morbific matter is not removed therefrom, soon proves inevitably fatal, by a general disorganization of the vital fluid. Especially, is this sooner or later the case, if the kidneys have been slow to perform their duty, or fail to exercise their compensating function.

Here the whole mass of the blood is contaminated by a specific poison or virus, and nature attempts to eliminate it through the medium of the skin, as is seen in the extensive eruption which covers the body. The supplementary duty of assisting in the removal of this specific virus from the blood is added to the already-increased duty of the kidneys, and in the discharges from these organs may we look for a solution of the nature of the complaint.

Now, respecting the character of the urine in scarlatina, Wilson says: "At the commencement of the disease, while there is considerable fever, the urine is of a deep, dark-red color, and possesses all the properties of inflammatory urine."

The blood in scarlet fever, be it remembered, is of the inflammatory type, containing an excess of fibrin. Also, that in children the urine is always less colored than in adults, and its color in this disease will be less dark; hence, the necessity of noting the age of the patient before proceeding to an examination of the specimen, for reasons mentioned in preceding pages.

The urine in this fever always has an acid reaction, is of a dark-red or brown color, soon becomes turbid, and often, says Schonlein, "resembles badly-fermented beer." At the commencement of the crisis the urine becomes clearer and forms a pulverulent sediment, consisting of uric acid.

These are the general characteristics of the urine in the regular form and course of scarlatina; but when, from imperfect elimination of the morbid matter from the blood, anasarca sets in, which is a very common sequel to this disease, a different set of characteristics appear in the urine.

Dr. Blackall says of dropsy following scarlatina, case four, in his work, that the urine was found so over-loaded as to resemble serum of the blood, three or four times diluted with water. It was high-colored, and contained a bloody sediment.

The bloody sediment observed in the urine in dropsy, after scarlatina, indicates how nearly it is allied to true hemorrhage. That it does actually derive its color from the presence of red blood, is evinced by its appearance, which can hardly deceive; by its being so speedily deposited, and by its remaining undissolved when heat is again applied.

Von Rosenstein states that the urine is mixed with blood in some patients, when the body swells after scarlatina.

Berserius describes it as, *turbida et fusca, et quandoque omnino suppressa;* and adds, in a note, that it had been found to resemble in color, water in which flesh had been washed.

Dr. Blackall gives quite a number of cases of dropsy following scarlatina, and seems to think that in all cases where the urine is coagulable, the tincture of digitalis is the *sine qua non*—the only remedy needed. He gives a number of cases in which the unloading of the urine by this medicine was effected in a short time, the improvement of the urinary discharge being simultaneous with the relief of the symptoms.

And, in some cases, wherein the medicines after having nearly brought about a natural state of the urine, and consequent return to health, by being discontinued too soon, "a return of the symptoms came on, the urine altering in the same proportion, and it was only by returning to the same medicine that a perfect recovery took place."

The tests for determining blood in the urine, as also that of albumen, both of which are of frequent occurrence in the sequela of this disease of infancy or childhood, will be found in another part of this work.

"In scarlatina," says Dr. Simon, "the urine is observed to be turbid at the time of desquamation; often, also, before the occurrence of the same on the outer cuticle, an extraordinary great quantity of the epithelium of the vesical mucous membrane is seen in it." "It is, therefore," says the same author, "to be admitted that the desquamation goes on also on the mucous membrane of the bladder, and if, as frequently appears to be the case, the scaling-off takes place earlier on the mucous membrane of the bladder than on the external skin, one may determine the commencement of the desquamation even by examining the urine."

With the high-colored urine of inflammation, and the excess of uric acid crystals, the age of the patient, the albuminous substance in the urine when the skin and kidneys do not remove it, and the epithelial scales in desquamation, anasarcous urine, etc., no one need for a moment be troubled in making out a diagnosis.

Treatment: Chlorate of potassa, diaphoretics and diuretics, with alkaline and stimulating baths, and washes or gargles of chloride of zinc in weak solution; when the throat is sore internally, place a mustard application to the throat.

We use a solution of chlorate of potassa internally during the whole progress of the disease, with tonics, when indicated. The sesquicarbonate of ammonia has lately been much extolled by medical writers, as having a specific therapeutic effect upon the special blood-poison of scarlet fever, neutralizing the so-called *materis morbi;* as also producing favorable results in the anasarca so frequently following, by its diaphoretic action in, "relieving the kidneys of the undue action they would have to perform in eliminating this blood-poison through the urine." Dose, from five to ten grains according to age; also use as a gargle with from two drachms to six ounces of water.

MEASLES.

THIS disease, although not denominated a fever, like the preceding one, yet it might lay claim to that appellation with equal propriety, as it is as similar in its relations, as to cause and effect, as is scarlet fever to the other idiopathic fevers. This is, however, essentially an eruptive disease, and as such, follows, with propriety, scarlatina.

Notwithstanding the similarity of the ushering in of these two disease, both being propagated by a peculiar poison, and both affecting the skin, in a peculiar manner, yet a marked difference in the blood and secretions is manifest, both during the disease, as also in the sequela.

Like scarlatina, this is essentially a disease of childhood, rarely affecting persons of adult age, and seems to be caused by the introduction into the blood, of a virus peculiar to itself, and is both epidemic and infectious.

It may be regarded as a disease of an inflammatory type, from the blood always exhibiting an excess of fibrin, and also, of a catarrhal character, from the urine, in the forming stage, always exhibiting the fresh-beer-like bubbles of carbonic acid.

Schonlein considers measles to be the most-highly-developed form of catarrhal disease, and says that the urine changes with the varying stages of the disorder. In most cases it more or less resembles the inflammatory type; it is red (as in inflammatory measles); acid; and sometimes turbid, as in gastric measles; or deposits a mucous sediment during the course of the morning, (as in catarrhal measles).

Becquerel states, as the result of his observations, that the urine is generally inflammatory at the commencement of the febrile period; that it becomes very dark, and of high specific gravity, and frequently deposits a sediment of uric acid.

During the eruptive period, the character of the urine changes. If the eruption is slight, and there is not much fever, it resumes the normal standard; if the contrary is the case, the urine retains the inflammatory appearance. And during the period of desquamation and convalescence, the urine either returns at once to the normal state, or continues turbid and sedimentary for some time; or becomes pale, clear, and anemic, according to the termination of the disease. The urine never becomes albuminous, like in scarlet fever, nor is measles likely to terminate in dropsy.

The above characteristics indicate pretty well the appearance of the urine and its different stages, according to the different stages of the disease in question.

The urine will always be found acid to test-paper, of high specific gravity; highly colored, and to contain crystals of uric acid.

With the knowledge of the age of the patient, to begin with, and urine with the above conspicuously-marked characteristics and qualities, we may very readily recognize the disease even where the patient is not present.

Treatment: Diaphoretics, stimulants and alkaline diuretics.

SMALL-POX.

THIS is a disease of an infectious and contagious character, and attacks alike the old and the young. It is also propagated by a peculiar poison in the blood, and manifests itself upon the tegumentary investment of the body, and therefore is an eruptive disease.

That this disease is one of the blood no one doubts, inoculation and vaccination having settled the point.

The blood, however, in this disease not only exhibits the usual marks of inflammation, but the peculiar matters which form the striking difference between it and every other disease, to-wit: the morbific virus itself; a purulent, mucous-like substance, not to be found in the blood in any other complaint.

With these facts before us, upon which we need not farther descant, we can easily anticipate the probable character of the urine. And, for the support of our own views, as we have thus done throughout this work, we will continue to supply evidence from every source of good authority, preferring quotations from these, to our own statements, whenever they contain the same opinions or facts.

It is remarked by Simon that the urinary secretion in variola undergoes changes, having relation to the various stages of the disease; that in the beginning, when the fever assumes the character of synocha, the urine is eliminated in quantity, and increased in specific gravity; that its color is deep and red, is frequently turbid, and often contains a small quantity of albumen.

We have often seen the urine, in the commencement of small-pox, and before there was any appearance of the eruption, so highly colored as to resemble bloody water.

In the eruptive stage, or when the eruption has more fully appeared, it becomes turbid and frequently throws down uric acid precipitates, either spontaneously or by the addition of nitric acid, and sometimes a little albumen, and even pus.

According to Schonlein, in the first stage of variola, the urine is of a reddish-brown tint; on the third or fourth day, a sweat of a peculiar and strong odor is observed, and the urine contains a turbid, and apparently purulent-mucous sediment, of an unpleasant odor.

During the suppurative stage of variola, Becquerel observed that the urine retained the synochal character as long as these symptoms continued; and in cases in which this fever persisted until death, the state of the urine always remained the same.

During the period of desquamation the urine is either normal or *anemic*, (generally the latter). In the putrid stage of variola the urine appears decomposed, ammoniacal, and not unfrequently of a dark-red color, from the presence of hematin. In the nervous form or stage of this disease the urine is pale and limpid.

To sum up, then, you will find the urine to be, in the forming stage, highly colored, bearing the general characters of inflammation, and perhaps reddened with blood. In the eruptive stage, it soon becomes turbid on exposure to air, and apparently contains a muco-purulent substance, and deposits uric acid crystals, and in the last stage, dark and muddy, with disagreeably strong odor.

Treatment: Allay the febrile excitement first, then give diaphoretics and stimulants, and finally, tonics. Keep the kidneys free with digitalis or other diuretics.

"An Indian remedy for variola" has been published in the Medical and Surgical Reporter, called "*Sarracenia purpurea*," or pitcher-plant. A wineglassful of the infusion of the root, it is said, brings out the eruption in those under the influence of the disease; a second or third dose, at intervals of four or six hours, causes the pustules to subside, apparently losing their vitality. If the patient is already covered with the eruption, a dose or two will dissipate it. The urine, from being scanty and high-colored, becomes pale and abundant; and in a few days the prominent symptoms subside, and no marks are left upon the skin. The remedy is used in the Indian camps to keep the antidote in the blood, and thereby act as a prophylactic.

ERYSIPELAS.

THIS is emphatically a disease of the blood, and one, the peculiarities of which are so striking and the danger in which is so great, as at once to convince us of the putrefactive tendencies of that fluid.

There can be no doubt as to its infectious character, as the experience of nearly all medical men has shown. That it is nearly allied in its nature to that state of the blood which produces puerperal fever in lying-in-women, is also fully established, from the frequent communication of that disease to patients of that class, as also, from children of puerperal patients being so frequently attacked with erysipelas.

An analysis of the blood shows an increase of the proportion of its fibrin, and a decrease in the number of its corpuscles. In this fact, will be found one reason for the tendency to the speedy and fatal terminations of this malady. The blood, in this disease, is poor in quality, and unable to resist the action or influence of the large quantity of foreign or morbific matters with which it is highly charged at the time; these are generally matters more properly belonging to 'the excretions, such as bile-pigment, nitrogen, etc., together sometimes with pus-globules, etc.

Andral and Gavarret found the blood in ordinary erysipelas so rich in fibrin, and its corpuscles so reduced in quantity, as to leave no doubt of its inflammatory character.

Schonlein states, that in this disease the serum is always tinged yellow by the coloring matter of the bile; that the proportion of serum to the clot is large; and that the consistence is inversely as its size. The quantity of fibrin generally rises to six or seven parts in one thousand of blood, three being the average in normality.

Simon observes, that in the early stage of erysipelas the urine puts on the inflammatory character. "It is frequently," Schonlein remarks, "loaded with bile-pigment, and is of a reddish-brown or very red color. At the urinary crisis, fawn-colored precipitates are deposited, and the urine becomes clear."

"Becquerel made two quantitative analyses of the urine of a man thirty-nine years of age, who had erysipelas of the face, and a good deal of fever, his pulse being one hundred and twelve. The urine of the first analysis was of a deep yellowish-red color, and clear; its specific gravity was one thousand and twenty-one. That of the second was so deeply colored as to appear almost black; it threw down a reddish sediment of uric acid, and had a specific gravity of one thousand and twenty-three.

In a woman, aged forty-five years, who had erysipelas of the face, whose pulse was one hundred and four and full, the urine was very scanty, of a dark brown color, strongly acid; it threw down a yellow sediment, spontaneously, and had a specific gravity of one thousand and twenty-three.

In five cases, in which the morning urine was daily examined with care, the characters of inflammation were present in a very high degree; the specific gravity varied from one thousand and twenty-one to one thousand and twenty-five. In four of these cases the urine threw down a reddish sediment, and in two, a little albumen was present. (Wilson).

In a majority of cases, at least, the urine will be of high specific gravity, acid to test-paper, will contain uric acid in abundance, and in bad cases, bile-pigment; in very bad cases, pus-globules, mixed with albumen, the urine changing from bright-red to dark-brown, black or muddy, in regular gradation.

Treatment: The treatment which seems to have been most favorably mentioned is that required for inflammatory diseases, in the first place; give stimulants and diuretics in the second, and tonics and stimulants in the third, with emollient poultices.

SHINGLES—(HERPES ZOSTER).

THIS is an eruptive disease, in which there is considerable of a change manifested in the blood, its general properties being so far altered from the natural standard as to produce considerable alteration in the character of the urine, according to the following authorities:

"The urinary secretion in herpes zoster," says Wilson, "has been made the subject of chemical examination by Muller.

"In one case, that of a boy eight years of age, the urine was abundant, faintly alkaline, pale-yellow, rather turbid; it rapidly became putrid, and deposited crystals of ammoniaco-magnesian phosphates; its specific gravity was from one thousand and fourteen to one thousand and fifteen.

"In the case of a young man aged nineteen years, the urine was clear, became turbid in the course of twelve hours, and deposited crystals of ammoniaco-magnesian phosphates; its specific gravity was one thousand and eighteen.

"In the case of a man thirty-one years of age, in whom there was slight fever, the urinary secretion was suppressed; that which was examined being the first that had passed for twenty-four hours. It was strongly alkaline, deposited a sediment of ammoniaco-magnesian phosphates, and urate of ammonia; its specific gravity was one thousand and twenty-eight."

The deductions resulting from the analyses of those three cases are," says the same author, "that there is: 1st, A marked increase of the chlorides and phosphates, and a corresponding diminution of the sulphates; 2d, An excess of hydrochlorate of ammonia; 3d, A large amount of fat; 4th, A diminution in the amount of uric acid, an increase only occurring when the disease is accompanied with fever—the presence of oxalate of lime may be suspected in these cases."

What more-marked indications of a disease would be required, to determine any case, than is given in the above? We need not add to the present, any experiments of our own; let the incredulous but examine for themselves, and they will learn that even the simplest disease makes its mark upon the blood, and correspondingly upon the urine.

Treatment: Sulphur, sulphuric and other acids, cream-tartar, vegetable diet, nitro-muriatic or sulphuric-acid baths; or sponge-baths, with sulphur and acids—the acids being sufficiently diluted.

PEMPHIGUS—(Vesicular Fever).

Even this extremely rare disease so fully makes its own mark upon the blood, or is so completely manifested therein, that the urine being secreted therefrom has its well-marked difference exhibited also. For the facts in the case, we present the opinions of learned men, where their views will coincide with our own experience.

The urine analyzed by Heller, in a case of severe pemphigus, which proved fatal, the patient being a woman forty years of age, was acid, and its specific gravity one thousand and seventeen ; it deposited a light, cloudy sediment of mucus, with fat-globules, urate of ammonia and epithelial scales. Of the fixed salts the earthy phosphates were normal, the sulphates much increased, and the chloride of sodium proportionably diminished. The urea was considerably above the normal average.

In the case of a little boy, affected with acute pemphigus, Dr. M. Wilson found the urine painfully acid, of light color, depositing a light, flocculent cloud, containing minute crystals of oxalate of lime, and loaded with urea.

For the examination of these kinds of deposits, we can not too highly recommend the use of the microscope. In fact, it should be the constant accompaniment of all investigations, chemical or otherwise.

Treatment: Alkaline remedies, diuretic tonics, neutralizing mixtures and soda drinks.

SCORBUTUS.

THIS, although not generally ranked among eruptive diseases, yet because of the ecchymosed condition of the skin, it may lay some claim to a place here, closely following shingles, pemphigus, etc.

"The humoral pathologists," says Foltz, "in placing the primary seat of this disease in the changed condition of the blood, come nearest, in our opinion, to a correct view of the pathology of this disease."

The most careful analyses of the blood in this disease have shown that it is one of poverty of that fluid; that the amount of fibrin and corpuscles are always diminished, while the proportion of water is greater than in the healthy state.

Dr. Foltz, Surgeon of the United States Ship Bariton, says: "The whole condition of the scorbutic patients clearly proves that morbifiic actions are going on in the blood. The extensive ecchymosis, purpural and petechial, the rapid tendency to ulceration in all parts of the body; the diminished quantity, high color, and tendency to speedy decomposition of the urine, all evince that the blood is the seat of the pathological changes going on."

The blood in this disease has been found to contain an excess of the chlorides of soda and of potassa, amounting to one-fourth part above their normal standard proportion.

"Scurvy is a disease," says Dr. Beach, "which evidently arises from a depraved state of the blood." The pecliarity of this depraved blood being a diminished proportion of fibrin and corpuscles, and an excess of the chlorides of soda and potassa, we will necessarily look for a corresponding change in the constituents of the urine, and find such to be the case.

Dr. Fauvel, whose analyses of the blood in scorbutus seemed to differ from the above, although his cases were of quite aged persons, which might make some difference, asserts, in proof of his cases being genuine scorbutus, that "although there was a yellowish color of the entire skin, the urine giving no traces of coloring matter, proved it was not icterus, which it so much resembled," but true scorbutus. He was enabled to tell only by an examination of the urine, the true nature of the disease; or rather, the urinary secretiou was made the test-indication or proof-mark.

Protein is the basis of albumen, fibrin and casein, and these are the commencing or starting-points of all the tissues of the body. The products of highly-proteinized vegetables are identical with the constituents of the blood, and when these principles are wanting in the food, as in the case of seamen and others who are deprived of vegetables for any considerable length of time, scurvy or poverty of the blood is the consequence, because nutrition has been arrested. Hence the blood in scurvy is found to be deficient in fibrin and corpuscles, with an excess of water and the chlorides before mentioned. And this condition of the blood yields to the urinary secretion the constituents of which reveal the true nature of the disease, and at the same time, according to Foltz, "furnish us with the means to guard against it, and to effectually remove it, after it does occur.

The urine is scanty and high-colored, but the coloring matter is different entirely from that of the inflammatory type of diseases, being of a more dull-reddish, iron-rust color, the coloring matter being that of the iron carried out from the corpuscles.

It is alkaline to test-paper, and deposits on cooling, a sediment of ammoniaco-magnesian phosphates, and sometimes urate of ammonia, the product from the excess of soda and potassa in the blood.

When these are found in urine of heavy color, like that of iron-rust, without the usual marks of inflammation, we will have no difficulty in determining it to be a case of scorbutus, the remedies for which, the long-boasted specifics, lemon-juice, citric, tartaric, and acetic acids, together with acidulated drinks and vegetable diet, proves but the correctnes of the theory, they being effectual in nearly every instance, in changing the condition of the blood, and altering the constituents of the urine *pro rata*—the natural operation of a scientific principle, showing the harmony and simplicity of all the operations of nature.

SCROFULA.

THIS disease follows in order here, not because it may be classed as eruptive, but because of its being somewhat allied, the tendency of many eruptive ones being toward the development of the distinctive characteristic of scrofula.

The scrofulous diathesis, as it is called, declares itself under a variety of cutaneous eruptions on the scalp, the eyelids, the nose, ulcers, etc., but until the occurrence of tubercle, the sure sign of its development, it is not considered to have reached this class.

Scrofula is occasioned by a perverted or unhealthy development of the nutritive materials, destined for the repair of the body and the restoration of the blood, and consists, according to Simon, in a peculiarity of blood-development, under which the nascent blood tends to molecular death by superoxydation.

According to Dr. Williams, the tubercular diathesis is a degraded condition of the nutritive material, from which the old textures are removed and the new ones formed, and differs from fibrin or coagulable lymph, not in kind, but in degree of vitality and capacity of organization.

The tubercular diathesis has a wide range of development and may appear in the form of glandular affections, carries off bones, cancerous affections, affections of mucous surfaces, etc. However much these affections may seemingly differ, the general conditions of the blood are similar, a partial change only being produced by local development; hence, we have tubercles of the lungs, tubercles of the liver, bronchiæ, or cancerous tubercles; or scrofula, consumption, bronchitis, cancer, etc. Each of these conditions may be developed by the same surrounding influences, as that which would affect the lungs in one person might only affect the bronchiæ in another, and so on.

The term *anemia* is applied to that condition of the system in which the predominant character is a deficiency of blood; and in scrofula there is not only a deficiency in quantity, especially of the red globules, but there is generally a remarkable deterioration in quality. The red particles of the blood being always defective in scrofula, there is at the same time a deficiency of the fibrinous element.

"In scrofulous affections," says Simon, "the blood is deficient in solid constituents, especially in fibrin and corpuscles; the primary causes being probably due to a deficient formation of chyle, and to the influence of a moist, unhealthy atmosphere."

When examined under the microscope, some of the corpuscles of the blood appear devoid of color at the edges only, and some are entirely colorless. It is also found to be deficient in the normal quantity of salts.

And, accordingly, we find the urine to be generally below the natural standard in color, and above it in fibrinous characteristics. It also contains more of the water and some of the salts, especially of the phosphates of lime and soda, which are always found to be in excess in the urine of this disease.

Therefore, in the examination of the urine of a scrofulous patient, having determined from the deficiency of color and normality of fibrin, the anemic state of the blood, the excess of lime and soda in the urine, together with the abnormal ingredients of the tubercles, or scrofulous pus itself, which, in such cases, is imparted to the urine, and which the microscope will fully reveal, we will have no difficulty in deciding as to the scrofulous character of any such specimen of urine.

By a microscopic examination of the urine containing these tubercular deposits, we are not only enabled to determine the general scrofulous condition of the system, by their cretaceous, calcareous, carbonaceous, pigmentary, nucleated, or cancerous formation, but to describe the particular development before mentioned, to-wit: whether from the lungs, liver, bones, or whatever tissues.

"In the cretaceous and calcareous forms of tubercle," says Carpenter, "the corpuscles and granules are mixed with gritty, saline particles, of an irregular form and size. A tubercle from the lungs and bronchial glands is pure carbon or charcoal, and chemically indestructible." "The cells of a cancer are large," says the same author, "transparent and distinctly nucleated, and consequently, easily distinguished from the small, non-nucleated corpuscles of a tubercle."

Crystals of cholesterine are sometimes found in the cretaceous and cheesy varieties of tubercle, also irregular black masses of pigmentary matter.

For a more minute and extended account of the microscopic examination of the different kinds of tuberculosis, refer to Smith's Appendix to "Carpenter on the Microscope." That author says: "By the aid of the microscope, the impositions so frequently practiced upon the physician, have been often detected. One of the most frequent of these attempts at deception consists in mixing various substances with the urine."

Scrofulous tubercles are generally found to be molecular and the corpuscles are larger and rounder than tubercles from any other part, and therefore different from those of consumption, the bone variety, etc.

"From some form of tubercle," says Newton, "scarcely an organ of the body is wholly exempt, and since the blood is the true source of the tubercular deposit, this is not surprising."

There is a peculiar substance resembling mucus, called cystine, sometimes apparent in the urine of scrofulous patients, especially those of a hereditary character. "The pathological character of this substance," says Bird, "is essentially connected with the excessive elimination of sulphur, every ounce of this substance containing more than two drachms of this element." We have seen a six-ounce bottle of urine let fall by repose, a sediment of cystine of the hight of half an inch.

It is believed to be peculiar to scrofulous cases, and especially to those of a hereditary character. "In one family alone," says the same author, "several members were at the same time affected with cystine; and one instance exists where it can be traced with tolerable certainty through three generations.

The color of the urine is generally yellow, sometimes changing to green or apple-green. Its odor is that of sweet-brier, or sometimes like that of putrid cabbage. On the evaporation of the urine, it crystalizes in the form of six-sided laminæ, but sometimes under the microscope resembles little white-rosettes. (See Chart, Fig. 9).

Treatment: Alteratives combined with tonics.

CONSUMPTION.

Of all the diseases afflicting the human family, the prevalence and fatality of consumption demands the greatest study and attention of the physician. The mortality of this disease has been set down at from one-fourth to one-fifth of the human race; or, "out of a population of twelve millions, sixty thousand die annually. (Beach).

There is in this disease an impaired physical condition of the lungs, and a deposition, in the air-cells, of a cheese-like substance, of a yellowish-white color, the particles of which aggregate and form masses of variable size, called tubercle, which finally soften into a liquid-like pus. This matter is taken up by the absorbents, enters the circulation of the blood, and particles of it are carried out of the system through the medium of the kidneys, and may be detected in the urine by microscopic and other modes of examination.

"The nature of this disease," says one writer, "is unquestionably scrofulous, as we see it transmitted from parents to offspring, with almost the same regularity as other legacies."

The blood in this disease is always unusually serous, its vitality therefore may be regarded as of a lower grade than natural, and is called anemic, while at the same time it may be thick and sizy, "containing the elementary principles of tubercle," says Beach.

While there is an excess of water in the blood, as also this thick, sizy matter, there is at the same time, a deficiency of red corpuscles and an excess of the fibrinous element. The average natural proportion of red particles in healthy blood, it will be remembered, is about one hundred and twenty-seven parts in one thousand, and in this disease it is sometimes reduced to as low as twenty-seven parts.

While this seems to be the case in scrofula also, we will have to ascertain the other peculiarities which distinguish it from that disease, in order to arrive at its peculiar development. This is to be done by obtaining a knowledge of the character of the discharges proceeding from the locality upon which the disease has spent itself.

The urine, then, coming from the blood in this watery state, of a sizy, yellowish and creamy condition, will partake more or less of these substances and characteristics. Not only will it partake more or less of these substances, but of those which are taken in as food, drink, etc., destined for the formation of blood-corpuscles, and which, by the peculiarity of the disease, are continually refused, as well as those which had already been formed in the vigor of life and are now being cast off. These substances will all be imparted to the urine, in the proportion of their abnormality in the blood, in the one case, and the refusal of their acceptance in another.

Therefore, as is always the case in this disease, the urine will be high-colored during the process of the removal of the red corpuscles, the color being of a peculiar crimson or cherry-red, and containing at the same time a portion of the sizy substance, which readily coagulates, forming a table in the center of the vessel, when a few drops of nitric acid are added.

When the red corpuscles of the blood are nearly exhausted, by the progress of the disease, the patient all the while becoming more exhausted also, the color of the urine becomes lighter, more straw-colored, and finally clear, as extreme debility comes on; but all the while maintains the peculiar character of coagulation in the center, as before mentioned.

By this change in the color of the urine, and an increase of the coagulable substance in the tabular form mentioned, a progression of the disease is certainly indicated. When, on the contrary, if the urine assumes a more natural amber-color, with less, or a thinner table of coagula in the center, on application of the nitric-acid test, improvement in the case may be decided on.

Let me here remark, that this improvement has been more uniformly effected by the administration of the sirup of iodide of iron, than any medicine we have ever given.

The chemical composition of the urine in consumption, shows an excess of the coloring matters, and of the phosphate of lime and soda, the result of non-assimilation and non-retention of those substances in the system, and traces of the peculiar tuberculous matter, to-wit: the "chemically indestructible matter of carbon, or charcoal." Tubercle from any other part of the system, be it remembered, is more easily destroyed by the action of alcohol or the mineral acids. They all vary in chemical composition, according to the location from which they are derived.

For microscopic examination, set the vessel of urine aside to cool. If a cloud be deposited in the bottom, pour off the supernatant fluid, place a drop of the deposit on a glass plate, evaporate to dryness and place it under the object-glass, when the nucleated, granulated, calcareous or carbonaceous appearances will be easily detected; and these give evidence of the nature of the disease, by the location from whence derived. No disease is more easily determined by an examination of the urine, than is consumption.

Before concluding we will present the views of A. P. Ducher, M. D., on *phthisis pulmonalis:*

Let us now briefly consider whether there is any peculiarity of urine characteristic of phthisis. If so, what does it consist in? We know perfectly well, that the great work of the kidneys is to separate from the blood certain nitrogeneous materials which are no longer fit for circulation. From the investigations of several experimentors, it appears that a certain relative proportion of uric acid, is essential to a healthy state. Now in phthisis, this is disturbed, as the tables of the different experimentors will show: We believe it may be stated, as a general rule, that whenever from any cause, rapid waste of the system is proceeding, an excess of uric acid will be found in the urine. Such is always the case in phthisis, when it is not complicated with kidney disease.

But the chief characteristic of the urine in this disease, is the production called *eurerythrin.* This is a beautiful carmine precipitate, and is easily detected by the addition of ammonia to the urine that contains it; it is seldom found in any other disease. We may generally suspect suppurative cavities in the lungs when this sediment appears in the urine.

In the first stages of phthisis, the specific gravity of the urine is very little below the natural standard, but in the last stage, it assumes what may be called *anemic* urine, presenting a pale aspect, without sediment and of very low specific gravity. If these conditions of the urine be kept constantly in view, they will, without doubt, at times, be of great service in making out a clear diagonisis of this disease.

Dr. Churchhill, who obtained notoriety by his discovery of the cure of consumption, by hypophosphites of iron, lime, and of soda, says: "All we can reasonably ask is, *that the treatment should dissipate the diathesis,*" which is very good. But furnishing the food to a disordered stomach will not alone feed the body; and in the treatment of this disease there is as much need of preparing the sanguiferous system for the reception of the blood-food, by the removal of the cause, as there is in fitting the dyspeptic stomach for digestion.

DROPSY.

The term dropsy signifies a morbid accumulation of water or serous fluid in the cellular tissue or serous cavities. It depends upon a loss of vitality of the nervous energy, of the capillary exhalents of the blood-vessels, together with a deficiency of what is supposed to be iron, and some other constituents of the blood.

Like scrofula, consumption, and other anemic diseases, it is one of *debility* of the blood, the most general cause of which appears to be a special difficulty existing either in a functional or organic affection of the kidneys. "This," says Dr. Beach, "has always appeared to me to be the great first-cause of dropsical effusion, and it has of late years been confirmed by the researches of Dr. Bright, of London."

Now, if it be even true that the most general cause of dropsy arises from a special difficulty in the kidneys, there are *varieties* of this disease that are complicated at least, with affections of other organs, such as of the liver, spleen, stomach, heart, and even of the generative organs, and to determine these complications is a desideratum—to be obtained many times only by an examination of the urine.

When the urine of dropsy throws down copious lateritious or fawn-colored sediments, as in disease of the liver, or is colored with bile, as in jaundice, or appears of a dark-red or brown color, without sediment, as when complicated with disorganization of the female generative organs; or when it contains blood, as in some pneumonic affections, these various peculiarities added to those which arise solely from disease of the kidneys, or from that which is characterized as dropsical urine, will enable us to determine, if not the cause, the complication, at least, and hence will give a clue to the proper treatment.

By whatever combined influence dropsy may ensue, the blood first partakes of these influences, and imparts the characteristic particles to the urine.

"The blood," says Beach, "is unhealthy in dropsy; serum or water, of course, greatly preponderates. Owing to the deficiency of albumen, the blood is pale-colored; it sometimes contains urea in large quantities. Albumen, fibrin and the red particles, which constitute the great bulk of the matters existing in the blood, are never met with in healthy urine; but in some varieties of dropsy and other diseases, the urine not only contains the serum of the blood, but the fibrin and red particles likewise pass through the kidneys unchanged."

"It appears, therefore, from what has been said, that the contents of the urine and of the dropsical fluid in this disease, bear to each other a certain definite relation. For example: the water which ought to be discharged in the form of urine, passes out in the form of dropsical fluid, and is not retained in the blood (but is found in the tissues or cavities), and the albumen, which is deficient in the blood, is found largely in the urine, scarcely at at all in the dropsical fluid, while the salts which are wanting in the urine, are found entirely in the dropsical fluid, none of those which ought to have been secreted with the urine being retained in the blood."

In the elegant extract above, we have the whole matter before us, in which we can see at a glance the pathology of the disease, to-wit: disease of the kidneys, the anemic condition of the blood (deficiency of albumen, fibrin and iron), with a large increase of albumen, etc., in the urine; as also may be seen the symptomotology of the disease, and a very good clue to its treatment is obtained.

There are several stages, however, of this disease to be recognized clearly, in the examination of the urine, in each of which, it changes in color, as well as chemical qualities, all partaking, however, of the same general characteristics before mentioned.

The first of these is that of congestion, in which the urine, loaded with albumen, etc., will be reddish, highly colored with the red particles of blood, and acid to test-paper.

In the second or more chronic stage, the urine will be paler, of a light-yellow color, and turbid, depositing more or less a brownish sediment, with acid.

In the third stage, the urine will be of a deep-red color, containing more or less of blood, the consequence of weakened blood-vessels, and will deposit a red or reddish-brown sediment.

"The urine in this disease," says Beach, "is almost always very acid, which fact must be kept in view in making out a diagnosis."

The urine, whether in dropsy of the head, chest, or abdomen, will exhibit the same general characteristics, modified or altered only by the local cause or affection coexisting.

Albumen will be found to enter largely into the ingredients composing urine from a dropsical patient, the quantity corresponding with the extent of the disease. This may be determined in the usual way, to-wit: producing coagulation by nitric acid and the application of heat, both of which are necessary to a surety, the heat giving the urine the appearance of curdled-milk.

The blood in the urine may nearly always be known by the dingy color it imparts to that fluid, or by tests, or microscopic examination.

If the dropsy does not proceed from a faulty action of the kidneys, whereby the blood has thus become deranged in its constituents, or if it is connected with the deranged condition of other organs, another condition of urine will be added.

"If the sediment in the urine is of a pink-color, it will always denote a scirrhus state of the liver," says Cullen. "In the diseased liver, the pink-deposits are almost constantly present in the urine," says Bird, "and we think we have received some assistance in the diagnosis between dropsy depending upon hepatic and peritoneal disease, in the presence of pink-deposits in the former, and their general absence in the latter."

The urine in dropsy will also be found deficient in the quantity of salts common to the healthy standard of that fluid, the consequence, as before observed, of their deposit with the dropsical fluid in the cavities.

These salts, chloride of sodium, potassa, lime, magnesia, etc., in health, generally constitute about fourteen parts in one thousand of that fluid, while in this disease almost their entire absence is noted; this state will be discovered most readily by the microscope. The evaporation of a small quantity of urine on a slip of glass, placed under an object glass, will readily enable one to tell, by comparison, very nearly the proportion of chloride of sodium, by the quantity of dagger-shaped or crosslet-crystals, etc., and of the other salts.

Treatment: In all cases of dropsy of the chest or abdomen, the warm, alkaline-bath, with bandage or compression of the swelled parts, seem to favor the evacuation of the fluid, and assist the action of the medicine in removing the accumulated salts, and in retaining the albuminous substance by supporting the weakened blood-vessels of the part.

Where albuminous urine is strongly presented, digitalis, in decoction, seems to be indispensable, and will frequently relieve the urgent symptoms in a short time. The use of this should be followed by tonics and astringent diuretics, uva ursi, buchu, and pipsissiwa in tea or decoction. If disease of the liver is connected with it, podophyllin, leptandrin, and iodide of potassa should be used, with perhaps cream-tartar, elaterium, etc.

DROPSY—(After Scarlatina).

"The different conditions of the urinary discharge," says Blackall, "seem to indicate a corresponding difference in the constitutional affection to which they belong; and we entertain hopes, that hereafter, and under a more accumulated experience, they may be found important guides to practice."

This is precisely the object of this work, to collect the experience of others, which when added to that of our own, may be found to be a guide to practice. Hence our reason in part for such free use of the labors of others, in the many quotations to to be noticed throughout its pages.

In dropsy following scarlet fever, the blood is found to present the same general characteristics as in other dropsical swellings, to-wit: a deficiency of albumen, iron, or red corpuscles, etc., a preponderance of serum, and its pale-colored and anemic appearance generally.

The urine, therefore, will of course take on the usual characteristics of dropsy. The same excess of albumen and deficiency of salts being apparent in the same relative proportion, according to the extent of disease, bearing in mind the age of the patient.

In the treatment of this disease, the facts respecting the relative conditions of the blood and urine going hand-in-hand, are only the more fully established.

Dr. Bird says: "The disappearance of albumen in the urine and the presence of uric acid, become valuable indications of convalescence;" while Dr. Blackall gives the most convincing proofs, that dropsy, after scarlatina, can be removed only by unloading the urine of the serum of the blood, or of its albumen, and says, that an improvement in the state of the urine will be among the first and most convincing signs of an improvement in the health of the patient.

His main dependence in the treatment of such dropsies, is, upon unloading the urine of its albuminous quality with digitalis; and if bloody sediment is at any time observed in the urine (which he thinks is generally hastened and encouraged by the use of mercury), he recommends the use of Peruvian-bark and wine, but in all cases to ascertain the exact character of the urinary discharge before employing medicine.

Our own experience and observation in this disease, which we believe to be often the effect of improper treatment and bad nursing of the fever, has led us to form a high estimate of diuretics generally, in connection with diaphoretics and sudorifics. If the treatment, indicated in this work for scarlet fever, be rightly adopted and carried out, there will scarcely ever be need for any afterward.

The warm-bath in these cases has a most beneficial effect, and digitalis is a most excellent remedy in connection. The sesquicarbonate of ammonia, as recommended in scarlet fever, may also be good. Dandelion, queen-of-the-meadow, horse-radish and juniper-berries, are among the vegetable remedies recommended; while acetate of potassa, nitrate of potassa, cream-tartar, etc., are many times used.

HYDROTHORAX.

In this disease, like the preceding one, there being the same general conditions of the blood, and correspondingly of the urine, there will not be a great or material difference to note in the case. Nor does it at all make any very material difference, since the condition of the blood and urine are the same or similar in each, the treatment, of course, should be similar; the same indications to fulfil and the results to be effected will be best accomplished by the same medicines.

There is, in fact, no very material difference in the character of the whole list of dropsical diseases, only in so far as they may be dependent upon different causes, or complicated with different other affections, and therefore give rise to additional symptoms; thus, dropsy may be complicated with diseases of the heart, liver, lungs, kidneys, etc., in all of which the peculiarities of urine belonging to these affections will be superadded to it in this disease.

There will be a difference in the external symptoms of dropsy, according to the locality of the disease, to wit: in hydrothorax there is difficulty of breathing, the consequence of invasion by the dropsical fluid upon the territory of the lungs; in hydrocephalus, a violent pain in the head, consequent upon pressure of the brain, and so on. But the only facts really worth knowing for the purpose of treatment, are whether the fluid occupies the cellular tissue, or serous cavities, and what is the primary cause of the affection. These may be determined, first, by the locality of the fluid; second, by the signs of the affection of the organs, as given in other places.

The complications, which are many times presented in this way, are the most difficult to determine, because of the obscurity which one throws around the other; this but increases the difficulty of a scientific treatment of an already difficult case. A correct knowledge of the peculiar cachectic state of the blood, however, to begin with, will very much enhance the prospects of a scientific treatment, and secure better chances of success.

Similar treatment will be required in this as in other dropsical cases, where there is no complication, and as it is not the province of this work to lay down definite and specific treatment of diseases, but rather to assist in diagnosing them, that which seems to be indicated merely will be mentioned, leaving the practitioner to select his own remedies and mode of applying them, in each particular case, according to his own good judgment.

That which carries off the dropsical fluid, and prevents its re-accumulation, will give relief the quickest. Digitalis, dandelion, queen-of-the-meadow, uva ursi, buchu, etc., with tonics and hydragogue cathartics, are indicated.

HYDROCEPHALUS.

THIS disease seldom attacks any but those of extreme youth; and while we have reason to believe that the same general characteristics govern it as those of other dropsies, owing to the difficulty of obtaining the urinary discharges from infant patients in a sufficient quantity and number of cases to fully test it, but little is positively known.

It is in dentition, however, that the foundation of this disease is most frequently laid, or it frequently follows as a sequela, like anasarca after scarlatina, and like this latter, the improper or unsuccessful treatment of the one, most frequently follows the other.

Dr. Simon says, that as long as the kidneys act freely there is little or no risk in the symptoms of mere dentition, however severe and distressing these may be; but if the urinary excretion is diminished or suspended, and this state of things is permitted to continue without relief, there is much risk of alarming cerebral symptoms quickly making their appearance. When the urinary secretion is scanty and deep-colored, the circulation seems to be both oppressed and excited; and the rapid, in some cases almost instantaneous mitigation, after a copious discharge of water, is well known to all experienced practitioners. We would remark that as there is generally the complication of dentition in this disease, the treatment should vary in accordance.

OVARIAN DROPSY.

We do not place this disease in the category of the dropsical ones from any resemblance it bears toward that class. In fact, it admits neither of the same explanation, nor submits to the same remedies as apply to diseases of the serous membranes. But, like hydrocele, and other hydropic diseases, is to be viewed more in the light of an enlarged viscus than in diffused accumulations in the cavities.

The blood in these cases is not generally represented by any very excessive deficiency of albumen, but rather a normality in that respect, and although somewhat anemic, there is scarcely ever any albumen to be discovered in the urine, unless the tumor has attained considerable size. Therefore, the urine is not found to be distinguished by scarcely any dropsical qualities at all; and in cases where the general health is not much impaired, very little change is discovered.

Schonlein, however, says that in ovarian dropsy the urine is sometimes very scanty, and contains a considerable quantity of albumen, which increases in quantity as the disease advances; therefore, when the affection is considerable, the qualities peculiar to that viscus, also, may be represented therein in connection with the very anemic urine which is then presented.

In whatever case, however, the treatment most beneficial seems to be that which meets best the indications of the urine. If anemic, tonics, iron and wine; if scrofulas or cachectic, alteratives and tonics; if dropsical, digitalis and astringents; cures are seldom performed by medicines, more seldom by operations.

CHLOROSIS.

THE blood in chlorosis is generally deficient in fibrin, never above the normal standard in quantity, and it is said not to be so firm in consistence as in health; and while the amount of fibrin may be normal even, the amount of corpuscles are always much diminished. And, as in most diseases, the physical character and chemical constituents of the blood change during their progressive development, and are different in the incipient and in the fully-developed stages. Thus, as the disease is more fully developed, the fibrin becomes more and more diminished, as also the blood-corpuscles, and with this great diminution of these constituents, there is a larger and larger augmentation of the proportion of serum, until it is no longer capable of supporting its function.

The urine of chlorotic persons is usually pale, of low specific gravity, and resembles the urine of persons who have lost a considerable quantity of blood. Beckerel applies the term anemic to this form of urine, and as in a majority of cases in which it occurs, there is either an absolute deficiency of blood, or a scarcity of a truly vital portion (the blood-corpuscles), the term would seem appropriate.

The urine in chlorosis has, however, other distinctive properties. According to Beckerel, "it is very poor in urea," and from the intimate connection subsisting between the action of the metamorphosis of the blood-curpuscles on the one hand, and the production of urea on the other, this is not surprising, the absolute and relative diminution of the urea thus plainly indicating the cause of the disease being a want of the production of blood-corpuscles.

In this disease there is a decided anemic condition of the blood, the red particles being reduced from the normal standard proportion five hundred per cent., in some cases.

"Certain it is," says Dr. Ashwell, "that chlorosis primarily depends upon a morbid condition of the blood, which secondarily affects the ovaries and arteries by retarding their growth.

This is supported by the fact, that in the blood of chlorotic patients there is an increased proportion of serum, with a marked diminution of crassamentum. And dissections of those who die with this disease, have generally shown the ovaries to be in a diseased or scirrhosed state.

"The morbid character of the blood in chlorosis," says Beach, "must be ranked among the most convincing proofs of the truth of humoral pathology."

There is many times a faulty state of the kidneys in connection with this disease, in which case the albumen of the blood is generally early and freely carried off. In the first stages, therefore, the urine may contain considerable quantities of that substance, and it becomes paler, and contains less and less of the coloring matters of the blood, also, until its weak and limpid-water appearance scarcely exhibits any traces of the general qualities of urine at all. When the urine exhibits this appearance, the proportionate quantity of serum in the blood has become so great that it is said that a drop of the blood will scarcely stain white linen, and is very similar to the blood of patients who have suffered excessive hemorrhage.

The peculiarly clear, watery appearance of the urine, derived from this state of the blood, gives a clue to the appropriate treatment of the disease, that is well worthy of our consideration; "a most grave error is too often committed," says one author, "by considering it a local, and not a constitutional disease; and ignorant practitioners, by the untimely use of drastic purgatives and emmenagogues, have yet farther reduced the already-enfeebled powers and facilitated the advent of pulmonary disease.

Treatment: Mild cordials with aperients; warm clothing, regular exercise, etc., with iron, hydrastis, nourishing diet, and saline-baths, with mustard hip-bath, etc.

The hypophosphites of lime, of iron, and of soda, if pure and perfectly prepared, might be of much benefit, if the stomach is in order to receive them. From the observations of Simon, as well as Arnold and Gavarret, the administration of ferruginous preparations are decidedly beneficial. The changes wrought in the blood thereby were truly surprising, the amount of solid constituents were increased nearly one-half, and the increase of hemato-globulin was likewise extraordinary, and the changes in the conditions of their patients kept pace with those changes of the blood. Small doses only of iron should be used, lest the digestive organs be interfered with.

DIABETES.

THE pathology of this disease is very obscure, yet, what ever may be its origin, there is no doubt but its first manifestations are in the altered condition of the blood. Like dropsy, however, which is in a measure its kindred disease, there is, generally, a deranged state of the kidneys with a deficient state of the urinary secretion.

In the remarks of Areteus, in his chapter on the cure of diabetes, "he commences," says Blackall, "by stating that diabetes is a species of dropsy, both in its cause and in its general nature, and differs from it only in the channel by which the humors pass out of the circulation; that in dropsy they are deposited in the cavities of the body, in diabetes are carried off by the kidneys and bladder." He adds, that if relief is obtained in dropsy, it is by that channel, namely, the urinary organs; but that this relief consists rather in the solution of the cause, than in the mere removal of the burthen.

We all know that there are many states of the system in which, if the urine is increased in quantity, the body wastes, if lessened, it swells; the one, therefore, constitutes a diabetic state, the other, a tendency to dropsy, and both are from a similar cachectic state of the blood.

Much has been written and many ingenious theories have been devised to explain the manner in which the saccharine matters found in the urine are formed, and the process by which these elements are removed from the blood. Without stopping to present these, we will proceed to the condition of the urine.

The first and most prominent symptom in this disease is generally perceived by the discharge of a most extraordinary quantity of urine, sometimes averaging ten quarts in a day, and some well-authenticated cases are on record, where from twenty to twenty-five quarts were discharged in twenty-four hours. Its weight, when the disease is fully established, exceeds that of the liquids drank.

"It is saturated," says Beach, "with a saccharine matter, and is very sweet to the taste; an ounce of sugar has been extracted from a pound of urine." This peculiarity alone is sufficient to detect the nature of the disease. A great variety of plans, however, for testing this kind of urine have been presented, most of which are sufficiently accurate for practical purposes.

The urine being thus excessive in quantity, largely saccharine in quality, will be acid to test-paper; it will generally be of a pale straw-color, and the odor that of new milk. When left standing for some time, it will become somewhat turbid, and soon undergo, spontaneously, what is called "alcoholic fermentation."

"The most convenient means of ascertaining the presence of saccharine matter in diabetic urine," says Christison, "is to add to it some yeast, which gives rise to vinous fermentation, a most delicate test, as it can detect one part of it in a thousand parts of urine." Every cubic inch of gas given off, nearly corresponds in round numbers with one grain of sugar.

"Another equally delicate test is the growth of torulae," (See Chart, Fig. 12). "These spores or fungoid vegetations make their appearance in the urine whenever saccharine matter is present in however minute proportion. Their true nature can be readily detected by the microscope."

"Boil the suspected urine in a test-tube, with an excess of liquor-potassa, and if it contain sugar, it will assume an orange-yellow, or brown or claret-color, in proportion to the quantity present." (Moore).

"If a drop of diabetic urine be diluted with ten drops, or even more of water, the presence of sugar will be shown by the above test." (Heller).

A great number of other tests are given by different authors, but enough has been said to enable anyone to decide with sufficient precision the nature of any case, which is the principle object of this work.

Before concluding this subject, we would remark, that although there is a seeming similarity between dropsy and diabetes, respecting the anemic condition of the blood, yet a very material difference exists; and from the very difference in the character of the urinary discharges, we must conclude that a different course of treatment will be requisite also. It seems that the kidneys are universally affected in this disease; not so in dropsy.

Treatment should be directed to the restoration of the kidneys; in connection—astringents, diuretics, tonics, and stimuants; animal diet as much as possible, with attention to the skin, with warm clothing, etc.

LIVER-COMPLAINT.

It is well known that the liver is one of the most active organs in the animal economy. It is supplied with blood from both the hepatic artery and the vena porta. And, in addition to the change made by its action upon the blood, in the separation of bile, it affects still further changes upon the blood by drawing from that fluid the sources of its own nutrition. As the entire structure of an organ must necessarily correspond with its functions, with every variety of internal organization, there will be a corresponding variation in the secretion; and any undue excitement of an organ increases its secretion, or its sedation diminishes it, so will we have it thrown upon the blood. In this disease there is a peculiar cachetic condition of the blood which gives rise to a long train of diversified symptoms. There is generally a derangement of the general or ordinary function of the liver, as well as structural affection; and this derangement changes the quantity and quality of bile, which it is the office of the liver to secrete, and thereby, the blood also is changed.

From what has been said in preceding pages on inflammation of the liver (which see), the great importance of this organ in the action of many diseases will be readily observed.

In the more acute form or stage of this disease, carbon is more abundantly eliminated from the kidneys than in this, the chronic form, and the urine will be more bright-colored in the former and more dingy in this latter stage. The liver itself becomes structurally enlarged and unfit to perform its duty, when imperfectly formed bile wlll be thrown into the circulation of the blood, become diffused throughout the general system, and taken up by the kidneys and discharged through the urine. The countenance becomes yellow, the eyes have a yellowish tinge, the whole skin becomes changed in color and appearance; and the urine takes on a brownish, or lastly, yellowish color, indicative of the presence of bile in that fluid, and on repose a brick-dust sediment is deposited.

The appearance of the urine as also those other symptoms, however, will vary according to the degree of deviation from the natural appearance, or a healthy standard. It will be small in quantity, high-colored, of high specific gravity, and the brick-dust sediment will be bile-pigment. "Gall-pigment in the urine" says Simon, "is constantly a sign of the liver being affected." The more severe the case, the deeper the color and the heavier the sediment formed in the urine. In extremely bad cases the urine, after having undergone the appearances above mentioned, becomes still more dark and heavy, amounting to a heavy brown color, which upon repose becomes still darker even, assuming a blackish, or almost muddy-ink appearance This last, however, like that in the last stage of jaundice, to which it is allied, may be regarded as a hopeless case, the blood having become so far dissolved as to be beyond the power of human agency to restore it.

The urine of chronic liver-complaint always contains another substance called purpurine. "The presence of an excess of this substance," says Bird, "is almost invariably connected with some functional or organic mischief, of the liver or spleen."

This substance is distinguished by not being affected in color or transparency by a boiling heat, and may be detected by adding a little hydrochloric acid to some of the urine previously warmed, when a color varying from lilac to purple immediately occurs.

"In the malignantly diseased, in the contracted, hobnail, or cirrhosed liver," says Bird, "the pink-deposits are almost constantly present in the urine." These pink-deposits present the flesh-colored appearance in the urine, and are a common accompaniment, in even slight derangement of the hepatic function.

Bile is always present in this disease, and may readily be detected by the usual test. A mere ocular inspection of the urine in liver-complaint, is nearly always sufficient to determine at least the nature of the case, and many times, to give a very correct opinion of the extent of the mischief also.

Treatment: Hepatics and deobstruents are indicated. Podophyllin leptandrin, and taraxacum are the principal remedies to be relied on, and may be combined to advantage. Alteratives are sometimes beneficial, sarsaparilla, guiaeum, stillingia, etc., and an irritating plaster over the region of the liver, with warm-baths, etc. An emetic, perhaps, to begin with, would be good treatment.

JAUNDICE.

FROM the fact that the preceding disease (liver-complaint) frequently manifests itself in the form of jaundice, or that icterus is most frequently the consequence of impeded passage of bile into the bowels, or that because of this impeded passage, the bile is taken up by the absorbents and thrown into the circulation of the blood, jaundice is made manifest, a very similar effect to that produced upon the urine in liver-complaint, is to be expected in the urine in this affection.

Accordingly, we find a great similarity existing in the urine in these two diseases, so far at least, as the biliary condition of that fluid is concerned. The coloring principle of the bile, at least, and in fact, the bile itself, so manifestly thrown into the circulation, thence out by the urine, makes every case of jaundice thereby so sufficiently marked as to enable anyone with the least experience, to diagnose the case correctly, upon a mere ocular inspection of the urine alone. In fact, no practitioner can have had much experience in this disease without being able, at a single glance of the urine, to determine at once the jaundiced condition of the patient; as also, to decide very nearly the precise stage of the disease.

In this, like in many other diseases, the state of the urine denotes the character of the affection long before it is manifested by any other symptom or more outward sign. The moment the blood begins to take on any of the constituents of bile, the kidneys begin to secrete it therefrom, and the urine to discharge it, more or less, according to the extent of the abnormality.

Dr. Simon says: " In jaundice, whether it be idiopathic or symptomatic, the urine contains bile-pigment, which shows itself in the peculiar color which it communicates to that fluid."

The color of icteric urine may vary from a saffron-yellow to a yellowish-brown, brownish-red, or blackish-brown. In acute icterus, accompanied by fever, Schonlein found the urine at first of a dark-red or brown color, from the presence of bile-pigment; it afterward became darker, and at last, as black as ink.

Scherer mentions a case of long-standing icterus, dependent, apparently, on chronic inflammation of the liver, in which the urine on emission was clear, yellow, and perfectly neutral, but after standing a few hours, became acid, and deposited a large amount of bile-pigment, in a yellowish-brown mass, and in the course of twenty-four hours the yellow color of the urine became converted into a blackish-green.

Prof. Eberle says he has not met with a single case of jaundice in which the urine did not acquire a bilious hue. And whatever may be the cause, all authors agree that there is a deposit of the coloring matter of the bile in the blood and tissues of the body, and that the same is observable in the urine.

There is seldom any inflammatory indications in jaundice, therefore the first trace of coloring matter, even in the onset of the disease, may be detected by the usual nitric-acid test.

In the more marked cases, the urine assumes a still more dark and bilious appearance from the greater excess of the same matter, and as the disease progresses to the more dangerous stage, the urine becomes more and more dark, even to a dark-brown or blackish appearance, because of the large amount of black bile or bile-pigment being carried out through that medium. And so regularly is this course in the disease marked by the urinary secretion, that every stage can be pre-eminently diagnosed thereby. And so universally is the extremity of the disease so darkly and swarthily marked, both in the skin and the urine, that what was at first called yellow-jaundice, is now not inaptly called black jaundice, by the common people, from which they believe "there is no hope of recovery," and they are not very far wrong in their conclusions.

The urine in jaundice, however, generally differs somewhat from that of chronic liver-complaint in this, that the bile is of a more vitiated quality, and hence the urine is more turbid from the beginning, and in whatever stage, has a heavier and more muddy appearance than belongs to that of liver-complaint, and, as a general thing, gives out less of the carbonaceous matters of the liver itself, or of particles from that organ.

From these qualities we will have no difficulty in determining the difference, and in correctly diagnosing an uncomplicated case of jaundice, in whatever stage of the disease, by an examination of the urine alone.

The coloring matter in the urine in a bilious case, is generally called purpurine, which, by the addition of a little alcohol to the deposit containing it, will produce a fine purple tincture, the purpurine being given up to the alcohol.

Treatment: First give an emetic of Lobelin, as this is a most relaxing agent, then the following powders, two or three times a day: Three grains of podophyllin, twelve leptandrin, and twenty-four Dover's powder—divided into twenty-four papers. Take at the same time half a teaspoonful of tincture sanguinaria in warm tea, before each meal, and one egg in brandy one hour after each meal.

Alkaline remedies are highly recommended; pills made of Castile-soap and leptandrin might be used advantageously.

DYSPEPSIA.

All the internal parts of the animal body are covered by a soft, velvety, and highly vascular coat, called the mucous membrane. These surfaces secrete a viscid, stringy, and often tough fluid, somewhat different in character, according to the part secreting it; and this secretion of mucus may be increased or diminished according as the part may be more or less affected; as also it may assume anomalous characters, which unfit it for the purpose intended. Especially may that be the case with the gastric juice or the secretion of mucus from the stomach.

The subject of digestion has attracted much attention, for a great number of years, and yet but little seems to be known with certainty, relative to the modification the food undergoes in the stomach and alimentary canal, but more especially the anomalous characters that are sometimes assumed in the processes of chylification and chymification.

"There can be no doubt," says Simon, "that there are anomalies in the process of chylification, in consequence of which an unsuitable chyle is prepared and conveyed to the blood, modified both in its quality and quantity; but with respect to the particulars of these anomalies we are still perfectly in the dark."

Theso anomalous processes mark the character and progress of indigestion, and take on the name of dyspepsia, in which there may be a derangement of the stomach manifested in a variety of ways, the imperfectly chylified food entering the circulating blood, not only fails to replenish the different organs of the body, but becomes in itself a source of irritation. As a general thing, however, there is a torpid or inactive state of the liver, or some derangement of the chylopoetic viscera in connection with it.

When the stomach becomes affected primarily, and without some derangement of the hepatic system in connection, it is seldom of long duration, but will readily yield to appropriate remedies, and soon recover its wonted tone and energy. In these cases we find a peculiar state of the urine, indicating not only the defective assimilation of food, but an affection of the stomach itself.

Without examining the particular changes in the blood in this disease, it will be enough for our purpose to describe those in the urine; and Dr. Bennet is "surprised that so little attention has been paid to the urinary secretion in dyspepsia, even by those pathologists who have written professedly on the subject; as the changes that take place in that secretion afford most valuable indications, not only for diagnosis but also for treatment and the regulation of the diet."

"As we have seen," continues Bennet, "if the lithates are too abundant to be held in solution by the warm urine, it is turbid from the first; if they are all dissolved by the urine while warm, but too abundant to be held in solution when cold, the urine becomes turbid as it cools. When the digestive and nutritive processes are very much impaired, these changes in the urine may be observed, whenever it is examined. If they are less deeply disordered, it is only two, three, or four hours after the ingestion of food (according to the length of time it takes to digest), that the urine contains the anomalous salts, and is turbid, or becomes so on cooling. If the digestion is still less affected, the lithates only appear in the urine, after the ingestion of animal substances, or an article of food of difficult digestion, etc.

From the above quotations or facts, it seems evident that in these instances the presence of the anomalous salts in the urine is entirely the result of depraved digestion.

The kidneys, in dyspepsia, owing to the imperfectly assimilated substances, as well as the depraved condition of the bile, throw out the effete matter in the shape of ammonia, triple phosphates, oxalate of lime, etc., but especially is this latter substance a constant ingredient in the urine in every case of indigestion, from whatever cause. The principle morbid product then in dyspepsia, to be found in the urine, will be oxalate of lime, the urate of ammonia, etc., being often only a concomitant, because of some other complication.

Golding Bird has given a number of illustrative cases of oxalate-of-lime deposits in the urine, and says that from the symptoms presented in cases of this disease, there is no difficulty in proving, to a demonstration, the positive and constant existence of serious functional derangement of the digestive organs, especially of the stomach, duodenum, liver, etc., when these deposits are found in the urine.

To examine urine, for the purpose of detecting the existence of the salts under consideration, allow a portion passed a few hours after a meal, to repose in a glass vessel; on cooling it will become turbid, and finally a dense deposit of urate of ammonia will take place. This, especially if accompanied with purpurine, will give evidence of derangement of the liver in connection. If there be dyspepsia only, there will be in the lower layers of the urine an opake deposition of a cloud of vesical-like mucus. Decant the upper portion, and pour the remainder into a watch-glass and gently warm it over a lamp, when a deposition of the crystals of oxalate of lime will be present; this may readily be examined with the microscope. (See Chart, Fig. 8).

"By allowing the urine to repose, and placing a drop of the lowest strata on a plate of glass, placing over it a fragment of thin glass or mica, and then submitting it to the microscope, the crystals diffused through the fluid will become beautifully distinct," says Bird.

Chemically, oxalate of lime is known by being insoluble in aqua-potassæ, insoluble in acetic acid, soluble in nitric acid, and convertible, at a red heat, into carbonate of lime.

There seems to have already been enough said upon this subject to enable anyone to determine a case of dyspepsia, with or without the the usual complications; but in conclusion, we will make the following quotations from "Bennet on the Uterus:"

"From what precedes, it must be obvious that the examination of the urine is calculated to be of great assistance in estimating the extent to which the uterine disease has reacted on digestion and nutrition; it is also a valuable mode of ascertaining, week by week, how far those functions have rallied under the means of treatment used. Owing to the intimate connection which exists between imperfect chylification and the presence of lithates, etc., in the urine, and the facility with which their presence may be ascertained, if the attention of the patient is directed to the urinary secretion, and the nature of the changes that take place is briefly explained to her, she is put in possession of a most simple and efficient means of regulating her diet, both as to quantity and quality.

"She soon learns that by noticing the urine, two, three, or four hours after the ingestion of food, she can tell whether the meal has been properly digested or otherwise, and thus becomes able to diminish or change her diet, as may be required. The information thus obtained is the more valuable, as a dyspeptic may not be apprised of the food she has taken not having been properly digested, by any other appreciable symptom.

"These remarks apply with equal force and truth to some of the most ordinary forms of dyspepsia, when existing without any uterine complication."

In all cases of oxalate of lime in the urine, the patients complain of well-marked dyspeptic feelings, with many times great nervous depression, emaciation and hypochondriacal impressions.

The treatment of dyspepsia should be varied according to the extent of the disease, and the nature of the complications, if any. We have prescribed the following with very good success in ordinary cases, to-wit:

One grain of eupatorin and poplin, half an hour before each

meal; Beach's Neutraling Cordial, seven ounces; tinctures capsicum, ginger and myrrh, three drachms.

Dose, from half to a table-spoonful half an hour before each meal, with one grain podophyllin and three grains Dover's powder, to be taken at bed-time.

NERVOUS DISEASES.

There is a large number of diseases classed under the above head, all of which partake more or less of the same general characteristics.

As it is our principal object to notice only the peculiarity of the urine in different diseases, to better enable practitioners to arrive at more correct conclusions respecting the nature of the maladies, we find ourselves particularly favored with contributions from the research of others, in this class of diseases. And it is in this class particularly, that some more definite mode is especially required, because of the general incapacity of patients laboring under hypochondria, hysteria, delirium, and other mental disorders, to give any very reliable or correct description of their symptoms or feelings.

It has long been known that certain emotions of the mind very materially affect the discharges from the kidneys; or, at least, that these certain changes in the character and quality of the urine take place under these certain emotions of the mind. And if it be true, that every thought emanating from the brain, involves the decay of a certain amount of cerebral substance, and that such influence of mental exertion on the metamorphosis of tissue, is exhibited in the composition of the urine, as has been abundantly proved by the interesting experiments of Dr. Hammond of the United States Army, it surely becomes a question of great importance to ascertain the character of such secretion in all nervous and mental diseases or excitements.

Dr. Hammond conducted a series of experiments upon his own person for the purpose of ascertaining the effect upon the urine, of increased or diminished mental exertion. Performing the experiments upon himself, made it easy for him to obtain standard amounts of the several constituents of the urine with which to compare the results; and to exhibit the manner of conducting these experiments we make the following extracts from his own pen. He says:

"I therefore reduced my food, mental labor, and physical exercise, to a system; appropriating eight hours to sleep, three to bodily exercise, seven to study and six to recreation, eating, and the performance of daily duties requiring but little mental or physical exertion.

"During the twenty-four hours, I consumed sixteen ounces of beef (broiled and roasted), twelve ounces of bread, one ounce of butter, eight ounces of potatoes, and two drachms of common salt. In the same period I drank thirty-two ounces of water; no other food, solid or fluid, was taken into the system.

"Under the several conditions of food, exercise, etc., above specified, I examined on ten consecutive days, the total amount of urine excreted, during each period of twenty-four hours, taking a note of the amount of urine, uric acid, chlorine, phosphoric and sulphuric acids respectively."

The result of these analyses he gives in tabular form. The average products per day of this first course of experiments, sums up, the urine, in ounces and decimals of an ounce, the other products in grains and decimals of a grain; thus: urine, 36.55 ounces; urea, 670.62 grains; uric acid, 14.44 grains; chlorine, 154.80 grains; phosphoric acid, 43.66 grains; sulphuric acid, 38.47 grains.

He then, with a view of determining the influence of increased intellectual exertion, says:

"I doubled the number of hours appropriated to study, taking for this purpose, three hours from the number given to sleep, and four from that assigned to recreation, etc., making a total of fourteen hours of the twenty-four, during which the mind was intensely occupied. This system I continued for ten consecutive days.. The urine was examined in the same manner as before."

The result of these analyses, the second course of experiments, gives the following: urine, 43.50 ounces; urea, 748.33

grains; uric acid, 10.65 grains; chlorine, 172.72 grains; phosphoric acid, 61.15 grains; sulphuric acid, 49.15 grains.

The influence of diminished mental labor was next to be ascertained, and with a view to that end he proceeded and says:

"I, therefore, omitted studying entirely, and passed seven hours allotted to it in the standard experiments, in reading light literature, and otherwise beguiling the time in amusements requiring but little mental exertion. As previously, this was continued for ten consecutive days; the food, exercise, etc., remained unaltered."

The effects of this the third course of experiments upon the urine is given as follows: urine, 32.14 onnces; urea, 586.65 grains; uric acid, 17.12 grains; chlorine, 141.94 grains; phosphoric acid, 25.40 grains; sulphuric acid, 35.81 grains.

These results, he thinks, can not be doubted, and are too well marked to admit of any other conclusion than the following, which he briefly states thus: "1st. That increased mental exertion augments the quantity of urine. 2d. That by its influence, the urea, chlorine, phosphoric and sulphuric acids, are increased in quantity. 3d. That the uric acid, on the contrary, is very materially reduced in amount. 4th. That *diminished* intellectual exertion, produces effects directly contrary to all the above.

"Thus," he says, "the brain is seen to follow the same general law which governs the other structures of the body—increased use promotes increased decay, and the products of this decay, are, in like manner, removed from the system to make way for newer matter."

Intense mental labor, by accelerating the metamorphosis of the cerebral tissue, necessarily requires the removal of that tissue, and the above facts respecting the removal of these metamorphosed tissues by the urinary secretion, and their easy detection in that fluid, at once put us in possession of the data whereby we may judge of the nature and extent of the disease, as well as prescribe the appropriate remedy. It is from experiments like those of Dr. Hammond, and from the anatomical structure, chemical composition, pathological conditions, lesions, etc., and the changes produced in the blood by poisons and by disease, and the effects of these changes, conditions, lesions, etc., upon the urinary secretion, that this theory, of diagnosis by the urine is founded, and especially is it well supported in this class of diseases.

From the earliest time, the brain has been considered the established seat of the mind. The color and consistence of the whole "nervous system" is different from that of any other part of the body, its chemical composition varying from all other parts. And, according to Baron Haller, "a fifth-part of the whole blood in the body of man is sent to the brain." And, when we find the effect of certain poisons, as alcohol for instance, to be so universally the same upon the nervous system, and can trace it from the stomach into the blood, from the blood to the brain, and from the brain making its exit by the urine, as was done by Dr. Percy, have we not good reason to regard that secretion as an excellent medium for imparting a knowledge of the nature of the disease? Magendie adopted the view that all poisons are absorbed into the blood, however quick their action, and destroy life by contact with the vital organs.

"Arsenic has been discovered by Orfila, in the viscera and urine of those who have been poisoned by it, both during life and after death, and it would appear that the living, in the case of arsenical poisoning, are constantly eliminating the poison by the urine."

"The fact," says Taylor, "that arsenic may be detected in the blood and urine of a person who survives its effects, is a point of considerable importance in a medico-legal point of view." "An analysis of either of these fluids," says the same author, "may furnish evidence otherwise only satisfactorily obtained by a post-mortem examination of the body; and cases of criminal administration of arsenic to the living, which have hitherto escaped the hands of justice, owing to the want of chemical proof, may become as clearly established to the satisfaction of a jury, as if the poison had operated fatally, and had been found after death in the stomach."

The fact, then, that intense mental labor universally increases one set of constituents in the urine and decreases another set, and vice versa, in diminished mental labor; and that the constituents so found in excess in the urine, under high mental excitement, are the ingredients that are deficient in the urine of those who are deficient in mental capacity (as in the brain of idiots, for instance, in which there is an entire absence of the phosphoric element, and in their urine a deficiency of the same), as also the effect of alcoholic stimulants upon the blood, brain, and urine, and the hand-and-hand correspondence of the blood and urine, in cases of other poisons; these facts, we may say, furnish us a sufficient foundation for the investigation of nervous disorders by the examination of the urine, under the different stages of mental and nervous excitement that may arise.

APOPLEXY.

The functions of the brain and nervous system, like the functions of organs generally, depend on the supply of blood to the part, and the good or bad quality of that fluid, for their due or undue performance. A due proportion of the properties of the nervous substance of the brain itself must accompany the blood to that organ, to supply the wear and tear to which it is subject; also, the prompt removal of the old material must take place, to insure a healthy action of the functions of that organ.

"If blood does not circulate freely through a limb, the sensations are impaired and its power to act reduced; and if too freely, the sensibility is exalted. Over-exertion of the faculties, or excitement of the mind, is chiefly felt in the functions of its own organs, the brain and nervous system. Hence may arise," says Williams, "congestions of the brain, exhaustion of nervous power, apoplexy, epilepsy, palsy, etc.

This class of diseases may be divided into *excess* and *defect* of blood to the part. The one constituting those of the nature of congestion; the other, those rather of recession, or the deficit of stimulus to the nervous matter. The former constitutes the inflammatory, the other, the non-inflammatory type of nervous disorders; and in the one we find the urinary secretion giving evidence of the same character, the "reddish-tinged flocculi" being observable, with a reddish-gray sediment, the phosphoric acid and other phosphates, while in the other, these phosphoric deposits are observable in the clear, pale urine of defective blood.

The urine in apoplexy is always alkaline to test-paper, the specific gravity above the average, and contains a trace of the reddish-tinged flocculi, deposits of phosphoric acid and other phosphates; and if the disease is of a violent character, an excess of urea will be deposited.

This class of nervous diseases might be regarded as inflammato-nervous; and under this head might be named apoplexy, epilepsy, and catalepsy, when they are produced by an excess of blood to the brain; and here the urinary secretion is of particular moment in diagnosing or determining whether the results which characterize these diseases, have been brought about by excess of blood to the brain, or from other causes, as they are known many times to be.

The peculiar mark of inflammatory urine in the one case, contrasted with the absence of said mark from that secretion, when from other causes, will very readily enable one to decide.

The phosphates being in excess in this kind of urine, they will be dissolved by diluted hydrochloric acid, and found insoluble in liquor-ammonia or potassa.

Dr. Bird says: "One general law appears to govern the pathological development of these deposits, viz: that they always exist simultaneously with a depressed state of nervous energy," etc.

Treatment: As this disease so much resembles an attack of inflammation of the brain, and as a similar train of symptoms and pathological indications are apt to follow, the treatment required in the former would not be bad for the latter, after the first or most urgent symptoms are relieved by local applications, etc.

EPILEPSY.

When the above disease is marked, as it often is, by an undue determination of blood to the brain, the urinary secretion will give similar marks to that of the urine in apoplexy. And, in fact, so far as the treatment is concerned in such a case, during the epileptic fit, there should be but very little difference, as in both it should be directed at once to the removal of the tension upon the brain.

If, however, the difficulty should arise from other causes, or should there be no excess or determination of blood to the brain, ocular inspection of the urine will alone sometimes develop the fact, by exhibiting the general marks of disorder without the evidence of inflammatory action.

The urine will be of lighter color, of light specific gravity, and will contain more phosphoric acid than is contained in that of the sanguineous variety. In an uncomplicated case of epilepsy, the other functions of the body being all duly performed, the urine will be of the standard quality in every respect save the excess of phosphates.

Very ordinary mental emotions, however, will very materially and suddenly affect a change in the urinary secretion, anger, terror, surprise, joy, etc., all having a very decided effect thereon; therefore, to guard against these, the urine passed after a night's rest only, should be depended upon.

We have examined a great number of specimens of urine from epileptic patients, and in all those uncomplicated with other disorders, if the specimen used was taken shortly after the "falling fit," it was clear, pale and limpid, exhibiting evidences of nervous weakness; but that taken the following morning was generally hazy, indicating an excess of phosphates, and deposited the same on repose.

As time passes on from the attack, the urine becomes more normal in appearance, and in some cases soon assumes almost the natural amber-tint also, in proportion as the patient acquires his usual nervous energy.

The treatment should be such as would have a tendency to allay nervous excitement, and to stimulate the heart to a regular propulsion of appropriate blood to the brain.

If there is an undue proportion of blood thrown upon the brain, and an unequal circulation of that fluid, there is no remedy in our experience, equal to the English digitalis, given in decoction. It is, without doubt, the only remedy that will not disappoint the expectations; phosphate of iron, valerian, and scutellarin, may be used in connection with advantage, however.

If the excitement is occasioned by worms, uterine derangement, or any other cause, the treatment must be varied accordingly. We have cured a number of cases of epilepsy in the young male subject by appropriate treatment for the removal of worms in connection. Also, a number of cases of females, wherein uterine difficulties were presented, and which perhaps contributed much to the production of the fits. In this latter case we place great reliance in the virtue of the tincture of macrotys racemosa, in connection with our usual remedies.

DELIRIUM TREMENS.

"THE nervous system," says Williams, "is an especial subject of the disordering influence of intoxicating liquors." If taken in large quantities it induces cerebral excitement, and if continued too long, congestion of the brain, and in extreme cases the spinal marrow suffers; hence, apoplexy, palsy, phrenitis, etc., may result, in which case the same symptoms as are manifest in those diseases, would follow, and the same changed condition of the urine be presented, with the addition only of the alcoholic substance which may be detected therein.

The most disastrous consequences are exhibited by the habitual drunkard, who, in proportion as he indulges in liquor, becomes subject to delirium tremens, the "drunkard's disease." The alcoholic poison soon induces other affections, depraving the appetite and destroying the powers of digestion, which but adds to the nervous exhaustion—the patient drinking and starving, and finally dying from exhaustion of *inanition*.

The brain and nervous system having been kept in a continued state of excitement by the alcoholic stimulants, the blood all the while receiving no nourishment for replenishing the wear and tear of the system, until at last forbearance ceases to be a virtue, the nerves are made to succumb to exhaustion, and the mind to hallucination.

With this state of blood, its vital properties reduced to the last stage of poverty and want, even to the last drop of nourishment contained in it, what could we expect to find in the urine secreted therefrom? Clear as crystal-water almost will be the flow, the damning poison only, scents the system through.

Dr. Percy detected alcohol in the urine of a *mania-a-potu* subject, and Drs. Christison and Percy separated alcohol from the brain by distillation, as also from the blood of a habitual drunkard.

Delirium tremens is undoubtedly a disease of debility and exhaustion of the blood, which debility falls most heavily on the brain and nervous matters; and to which is added the active poisonous effect of the alcohol.

There will be no trouble in detecting this disease by an examination of the urine, after having once inspected and tested a specimen. In the early stage of the disease, and before the delirium has fully set in, the specific gravity of the urine may not be so light, nor yet its color so light, as it may still contain more or less of the tinge of inflammation, as also some of the unassimilated food of indigestion, as after the delirium has been developed. But in the wildness of delirium, the urine will be as clear as water, and contain a few small flakes of phosphoric acid, to be seen only by the microscope; and it will also have the oder of alcohol.

To detect the alcohol in the urine without the trouble of distillation, drop in the vessel containing it, a few grains of bichromate of potash, then add a few drops of the oil of vitriol. If alcohol be present, "even only in the proportion of a drop to half an ounce or ounce of water," says Dr. Thompson, "green oxyde of chrome will be set free, and the odor of aldehyde will be perceived.

This is a valuable test, and is worthy of confidence; distillation, however, very readily detects the alcohol.

As long as nervous exhaustion can be averted, by repeating the stimulus, and keeping up the strength and vivifying properties of the blood, the disease in its virulent form may be warded off; but when, in consequence of the complete exhaustion of these properties and the nervous matter, stimulants fail to arouse the system to its wonted action, or the organs to their regular functions, delirium, hallucination and insensibility take place.

Treatment: Rest, quietness of mind and body, and the administration of another form of stimulant in connection with tonics and diuretics. Give ammoniated valerian, as a counter-acting stimulant to the alcohol, with the tonics, hydrastin, lupulin, and iron, to bring up the state of the blood; and diuretics, as uva ursi, buchu, or digitalis, to carry off the alcohol.

Dr. J. C. B. Williams takes a rational view of the treatment of this disease and says: "The excretory functions are imperfectly performed, the urine is no longer freely secreted, and as long as the secretions are defective the nervous excitement continues; the leading object, therefore, should be to purify the system by means of an increase of the secretions, and especially by diuretic remedies." This would certainly be a justly scientific course of treatment of delirium tremens and deserves to be more frequently tried.

It is said that the *amanita muscaria*, a species of fungus, produces inebriation, and that the inhabitants of Northern Asia use it for this purpose. He who has eaten it will, in the course of twenty-four hours, have slept himself sober; when, if he takes a teacupful of his urine, he will again become intoxicated; and a party of drunkards, it is said, may keep up their debauch for an indefinite period of time by drinking the urine of each other—one only having eaten the fungus.

HYPOCHONDRIA.

THE above disease, together with some forms of insanity, generally arise from too great exertion of the mental faculties, or too long continued application of the mind upon the same subject or class of subjects, and consequent waste of nervous matter, together with want of blood-globules in the blood.

The chemical analyses of the brain of idiots, exhibits an almost entire absence of the phosphoric element, or phosphorus, according to the experiments of a number of authors; and in accordance with the deficiency of this constituent may the deficiency of mental power be estimated.

"Almost every act of the mind," says Dr. Carpenter, "is inseparably connected with material changes in the nervous system, and hence the more prolonged and energetic the operations of the mind, the greater is the waste of nervous matter, which is evinced by an increased amount of phosphates in the urine."

Here then we have the *data* for the measurement of the degrees of deviation from a healthy standard of the nervous system, which will enable us to calculate, with almost mathematical precision, the state of the mind even, by an examination of the urine.

The author above quoted farther says, that the peculiar ingredient of the nervous tissue, is a fatty acid, containg a very small proportion of azote, but united with a considerable amount of phosphorus. The amount of change which takes place in this or any other tissue, may be estimated in two ways: first, by the apperance, in the excretions, of its peculiar ingredients, set free by decomposition; and, second, by the demand set up for the material for its reformation.

Now it is well known, that, when the nervous system has been in unusual activity, there is a marked *increase* in the phosphoric deposits in the urine; and as the quantity of phosphorus in any other of the soft tissues, is very considerable, it is scarcely possible to attribute this liberation of phosphorus from the system to any other cause than the *waste* of nervous matter—that is, its decomposition, resulting from the discharge of its vital function.

It is evident to all investigators of this subject, that the urine in mental disorders, and especially those of a non-inflammatory and more chronic form, not only gives out more than its accustomed share of the nervous matter or phosphoric substance, but that the excess so given out, is coexistent with, and in the ratio of, the existing disease. We have then, only to discover the extent of deviation in this particular, from the healthy standard of urine, to enable us to determine, not only the character of the disease but its extent also.

Whenever, therefore, we find the urine to contain more than two andtwo-ten ths parts of phosphoric acid, in one thousand parts of urine, that being the average in health, all other parts of the urine being normal in quality and proportion, we can readily calculate the nature and extent of the disease. This kind of urine is generally paler than the natural, and if the disease is considerably extensive, is somewhat hazy in appearance. It is always alkaline to test-paper, and the phosphates may be easily detected by the addition of dilute hydrochloric acid, which dissolves it, or the addition of liquor-ammonia or pottassa which does not.

If there be complications, such as dyspepsia, for instance, these will be marked by their usual indications also, in the urine. The well-marked dyspepsia will always be shown by the oxalate-of-lime deposits in the urine, the tests for which ingredient, see Dyspepsia; also Chart, Fig. 8, for microscopic view. These will enable anyone to detect any variety of hypochondriacal or dyspeptic urine.

The treatment of hypochondria, should be such as would relieve the wear, tear, and excitement of the brain and nervous system, to prevent the disintegration of the nervous matter, and to supply the deficit that has been carried off.

Remove exciting and depressing influences, and substitute rest, quietude, but not solitude, with pleasing associations, and occasionally, even cheerful and joyous exultations.

Give bitter tonics, dydrastis, iron and wine, with the phosphates of lime, iron and soda; together with a bland, nutritious diet—beef-tea, soups and calves-feet jelly.

PARALYSIS.

This disease, although perhaps more nearly allied to apoplexy or diseases of the inflammato-nervous character, yet as it may sometimes be produced by enervation of the system, or as the loss or diminution of power of voluntary motion, are not always to be attributed to inflammatory action, it may be well to mention here that the urinary secretion is of the most vital importance in determining the facts and discriminating between them.

"Paralysis may arise," says Cooper, "in consequence of an attack of apoplexy. It may likewise be occasioned by anything that prevents the flow of the nervous power from the brain into the organs of motion. The long-continued application of sedatives, also, will produce palsy, as we find to be true of those whose occupations subject them to the constant handling of white lead, etc."

If, from an attack of apoplexy, or from any inflammatory condition of the blood, the urine will exhibit the usual marks in inflammation so well known. If from a lack of flow of nervous matter, that deficit will change the quality of the urine, in the reduction of one of its proportionate constituents. If from the introduction of a foreign substance into the circulation of the blood, as lead, for instance, that substance will be found in the eliminated urine.

Orfila detected lead in the urine of a female who had swallowed the acetate; "and during life it appears to be eliminated chiefly by the urine," says Taylor, on Poisons.

CHOREA—(St. Vitus' Dance).

This disease is generally of a non-inflammatory character, and belongs to the class of nervous disorders, of rather a convulsive kind. Some have thought it nearly allied to paralysis.

Various causes may contribute, however, to the production of this disagreeable disease, among which may be enumerated teething, worms, poisons, violent affections of the mind, as anger, fear, horror; as also, whatever produces general weakness of the system, and poorness of blood—*aglobulia,* or want of blood-globules.

From whatever cause the disease is propagated or the convulsive motions of the limbs may be produced, the characteristic marks of the urine will determine. If from teething, "the chief danger of dentition," says Simon, "is referrible to the vascular excitement of the brain." As long as the kidneys act freely there is little or no risk in the symptoms of mere dentition. When the urine is scanty and high-colored, the circulation seems to be both oppressed and excited. If from worms, the usual milky appearance of the urine will be perceptible. If from poisons, the substance or poisonous matter may generally be detected in the urine. And if from violent fits of passion, the urinary secretion will exhibit the purely nervous character, so well known.

The treatment must be varied in accordance with the cause of the affection, and in this lies the real value of the Uroscopian system—the detection of the true cause.

HYSTERIA.

THIS, the last of the nervous diseases, which we shall notice, though by no means the least, in point of vexation to practitioners, or frequency of attack, may be taken as the grand type of all non-inflammatory nervous disorders.

"Aglobulia, or want of blood-globules, has for its distinguishing mark," says Merchand, "a disturbance of the nervous system. Women have a smaller proportion of blood-globules than men, therefore they are more subject to nervous maladies; this smaller proportion of the blood-globules in them may be ascribed to their periodical hemorrhage."

Sydenham affirmed that two-thirds of the female world suffer from hysteric symptoms; and also conceived that there is no difference between hysteria and hypochondriasis.

It has been observed that nervous diseases have a tentency to subside after the critical age of women, and it is an equally observed fact that, after this period of life, the blood-globules acquire an increased proportion, owing to the cessation of menstruation.

It has been said, with some show of truth, that "chlorosis is the aglobulia of young girls and often of young men; hysteria, the aglobulia of females from twenty-five to fifty-five years of age, and hypochondriasis the aglobulia of grown men."

The medium or standard proportion of blood-globules being one hundred and twenty-seven parts in one thousand of blood, they are many times reduced in this disease to as low as forty-three parts, and in chlorosis sometimes as low as twenty-four parts in one thousand of blood, according to Andral.

It is a well-known fact, that "nervous persons" are generally spare and thin of flesh, or that those of a "nervous temperament are seldom or never fat."

The nervous matter is chiefly formed, according to Carpenter, "out of the same elements as those which would otherwise be employed for adipose tissue, but which the continual use of the nervous system prevents from being deposited, and which is carried out by the urine, as is observed in its phosphoric deposits."

That the urine in these disorders, and especially in hysteria, is of a peculiar character and quality, every physician of experience knows; even nurses themselves are well acquainted with the fact, that there is generally a large flow of clear, limpid urine after an attack of hysteria.

That there is an excess of water there is no doubt, the standard proportion in healthy urine being nine hundred and fifty-six parts in one thousand; in some cases the excess has been known to reach thirty-nine parts abov, emaking nine hundred and ninety-five parts, leaving only five parts in one thousand of urine to be made up of the other constituents, the principal of which were generally the phosphates.

The urine coming from blood which is almost destitute of globules, could not well be much else than water, where no other affection or contamination of blood is present. It will be clear, pale and transparent, scarcely a trace of anything to give it color or consistence.

"The experienced physician," says Aldridge, "can at once perceive, if the urine be pale and transparent, although strabismus, convulsions, delirium, heat of head, and bounding carotids should all display themselves, he can confidently say: 'There is no inflammation here; this is hysteria.'"

This condition of the urine, contrasted with the higher colored and more concentrated urine of the partially inflammatory, or the inflammato-nervous diseases, as apoplexy, etc., will be our chief guide in making up a diagnosis. Remember that there is, in the inflammato-nervous, the tinge of reddish flocculi, and the more natural pcoportion of water in the one, and the paleness and large *excess* of water in the other. The phosphoric deposits may be detected by the usual chemical and microscopical rules, where the unpracticed eye is unable to determine, or where the quantity is insignificantly small. Hysteria, however, may sometimes be the result of other disease, or at least, somewhat complicated with it, such as dyspepsia, uterine diseases, etc.; and in such cases the usual characteristic changes of the urine in these diseases will be apparent also.

The treatment indicated in all these kinds of disorders, in order to a perfect restoration, or to perform a complete cure, must be directed to the restoration of the blood to its normal standard, to increase the number of blood-globules and thereby reduce the watery excess in the urine. This is best accomplished by vegetable tonics and astringents, with generous diet, wine, iron, and stimulating bitters, with antispasmodics to calm the nervous system.

CHRONIC RHEUMATISM.

ALTHOUGH this disease sometimes follows that of the more acute form, or inflammatory type, still a very great difference is to be observed in the constituents of the blood as well as the urine.

It is now pretty generally conceded, and the experiments of Drs. Day, Garrod and others, show that it depends on the production in the system of an excess of lithic acid.

Rheumatism is especially liable to occur as an effect of cold, or check of perspiration, and as the perspiratory secretion contains lactic acid, the skin failing to excrete that substance, if the kidneys should at the same time fail to remove it, there must be an accumulation of it in the blood, which, acting as a ferment, with its kindred products, lithic acid and its compounds, contaminates that fluid.

Dr. Garrod readily detected lithic acid in the blood of rheumatic and gouty patients.

That there is an excess of uric acid, or lithic acid, in the blood of rheumatic patients there is not a shadow of doubt, and in this respect the acute form agrees also with the chronic; and that it acts upon the muscular and nervous systems, producing pain, contractions, etc., belonging to rheumatism, is also true. It appears to exist there after the form of some kinds of poison, or as an abnormal ingredient, which has accumulated in the system, and for a time is deposited there.

In the chronic form of this disease during the accumulation, and in fact almost during the existence of this substance in the blood and tissues of the body, to which it should be a stranger, there is little or no appearance of it in the excretions, to which that product more properly belongs.

This deranged or vitiated state of the blood is the first link in the chain that makes up a case of rheumatism, and the deranged condition of the organic system of nerves and muscles are merely secondary consequences. It is in vain, then, to attempt its cure by external applications.

The pathology of chronic rheumatism being the result of an accumulation of this acid in the blood, is farther proved by its non-elimination during the existence of the disease, and its characteristic appearance in the urine, upon the subsidence of the disease under appropriate treatment.

There is in this disease almost a total absence of lithic acid in the urine, until a crisis takes place, or by appropriate treatment that substance is eliminated by it, when almost immediate improvement of health is observed.

Dr. Prout says: "Whoever has attended much to urinary diseases must have remarked, that many individuals, subject to these derangements of health, seldom feel so well as when lithic acid deposits take place in the urine."

The same author says that lithic acid may be considered the *materis morbi* that is the cause of irritation in the constitution, when, if the kidneys can be made to secrete it in large quantities, the system will immediately be relieved, and thus an artificial crisis may be brought about by diuretic remedies.

The urine in the first stage will be entirely devoid of uric acid, but after the disease is fully set in, will begin to exhibit traces of that ingredient. It is always of lighter specific gravity than normal urine, until the crisis, whether from the efforts of nature alone, or the use of medicines, when it becomes heavier, and assumes a darker and muddy-like appearance. And, in this, as in many other diseases, "the heavier the sediment lies at the bottom, and the clearer the urine is that stands over it, the more decided is the crisis allowed to be; while the lighter the sediment floats, and the less disposition there is to a quick and perfect deposition, the more imperfect the crisis."

Of all the remedies for chronic rheumatism, colchicum seems to have been given the preference, as being the most certain to produce abundance of uric acid deposits in the urine, especially if given in combination with iodide of potassium and saline-waters. These increase the action of the kidneys, and drain the offending matters from the system.

Dr. Lewis says: "Colchicum causes an augmented discharge of this and other principles of the urine." "In slight cases of rheumatism," says Williams, "sudorifics may suffice, but colchicum and alkalies, opium, and iodide of potassium, more speedily and permanently remove the disease."

In old or long-continued cases of chronic rheumatism, concretions or thickenings take place in the fibrous, cartilaginous and white-tissues generally. These are owing to the deposit in them of soda and lime in combination with the lithic acid. It is in this combination that it is presumed to exist in the blood.

"Taking into account these two prominent facts, the excess of lithic acid found in the urine at the period of convalescence from rheumatism, and the subsequent deposit of soda and lime in the white-tissues," Dr. Buckler proposes to decompose the soda and lime in the blood, by the administration of phosphate of ammonia, and says that thickening of the white-tissues, of long standing, has disappeared under its use. Even in those cases where convalescence had already commenced, and the lithic acid was present in excess in the urine, it at once disappeared under the use of phosphate of ammonia.

It is thought by some, that to saturate the fluids of the body for a sufficient length of time, with the phosphate of ammonia, the calcareous substance must be dissolved.

GOUT.

THIS disease seems to be a species of rheumatism, being very similar in many respects. Dr. Black says, "the principal difference consists in this, that in rheumatism the local disease affects the larger joints, while the gout affects the smaller ones, the toes, etc."

Dr. Garrod has in several cases of gout detected lithate of soda in very appreciable quantities in the blood, while in the commencement of a fit of the gout there is a marked diminution of it in the urine. "While on the abatement of the attack, the lithic acid or its compounds," says Williams, "appears in increased quantity in the urine, that in the blood is diminished."

This is the state of the blood and also of the urine, and the relation they bear toward each other in chronic rheumatism, and that such is the view taken by most writers, is evident from the fact of the treatment laid down by most of them being generally the same in both disases.

"In gout," says Dr. Dick, "the urine previous to the attack and often during the severity of it, deposits uric acid, the urates of ammonia and soda, etc., and these salts, especially the urate of soda in combination with phosphate and carbonate of lime, when not sufficiently eliminated, are deposited in the cartilages and ligaments around joints, etc., and form what are called 'chalk-stones.'" This is after the same form as the concretions of rheumatism, and are to be removed by the same remedies. These concretions will be re-absorbed and removed in the form of uric acid in the urine, by a lengthened course of treatment, by colchicum, iodide of potassium, etc., or perhaps by the administration of the phosphate of ammonia, as recommended by Dr. Buckler, in rheumatism.

GRANULAR KIDNEY.

THIS disease is characterized by a morbid deposit in the substance of the kidney itself, changing the appearance, and structure even, of that organ. The kidney loses its usual firmness, changes its color, becomes more yellow, and is sometimes filled with an opake, white deposit, and finally the external becomes rough and scabrous, with projections innumerable, of a yellow, red or purplish color, and not much exceeding in size a large pin's head. It is the same disease as that called "*Morbus Brightii*," the principal and distinctive feature of which is, the presence of albumen in the urine with a deficit of urea. The secretion of urine is somewhat scanty, high-colored, sometimes tinged with blood, and very highly charged with albumen.

So universally has this latter substance (which is not an ingredient in healthy urine) been found in this state of the kidneys, that the disease has been called "albuminaria."

From the albuminous urine being so generally connected with a dropsical state of the system, it may readily be supposed that there is a tendency to that state in this disease, which we find to be the case. In fact, this disease is generally complicated with dropsy, and sometimes makes its appearance in that form, even before we have had any other indication or notice of the real difficulty; and whether or not, of all the secondary consequences of granular kidneys, dropsy is the most frequent.

To test the albuminous quality of the urine, if it becomes opake by heat, as also by the addition of nitric acid, both of which should be tried, it will be albuminous in extent according to the copiousness of the coagulation. The dingy-red appearance of the urine will lead to the suspicion of the existence of blood therein, when, by dipping a piece of white linen into the vessel containing it, if blood be present, it will be tinged with red; or, by the addition of alcohol, the urine will become lighter.

The readiest and most convenient, and likewise the most infallible mode for detecting blood in the urine, where great exactness is desirable, is by the microscope. Allow the urine to repose in a tall glass, take a drop from the bottom of the vessel, and place it under the object-glass. (See Chart, Fig. 11, C and D).

Treatment has been very ineffectual in the generality of cases of this disease. Strict confinement to the bed, with the administration of small doses of turpentine, balsamic preparations, with uva ursi, chimaphila umbellata, diosma crenata, trillium, conium. etc., are most likely to benefit.

SPINAL DISEASE.

This disease generally proceeds from an injury, and is characterized by a high state of nervous irritability, pain over the lumbar region, with irritability of temper, etc. And the state of the brain and nervous system is so manifestly and morbidly irritable, that the secretions, and especially the urinary, produce irritability of their parts also. This irritability is generally sufficient to give rise to suspicions, and may be traced to some recent or remote injury to some part of the spinal column, either from a wrench of the back, blow upon the spine, or some previous sprain.

The urine is generally more copious than usual, frequently pale, and of specific gravity below the average. The deposits are of the phosphoric kind, and almost exclusively of phosphate of lime. It is always alkaline to test-paper.

The fact of alkaline urine resulting from strains or blows on the back, was first noticed by Dr. Prout; and injuries to the loins have long been enumerated among the exciting causes of renal calculi.

"This alkaline state of the urine, and deposition of phosphates," says Bird, "is a pretty constant result of anything which depresses the nervous energy of the spinal marrow, whether the result of insidious disease of the spine, or the effect of sudden mechanical violence."

In injuries of the spine the urine is not only always alkaline, but the earthy phosphate is that of the phosphate of lime only. In other diseases of the nervous system and of the brain, there is generally the triple phosphates; and if it be connected with dyspepsia, the oxalate of lime will be present; for as injuries to spinal marrow and dyspepsia may both afflict the patient at the same time, so may we find in this case the resultant phosphate of the one, and the oxalate of the other, in the urine at the same time; and so in other complications of diseases. (For microscopic appearance, see Chart).

By the presence then of phosphate of lime in the urine, are we enabled to confidently diagnose a case of injury to the spine; and even although the injury may have been received during the existence of another disease, this mark of the urine will lead to the suspicion at last, that it is complicated in this way.

An amusing story has often been related of an incident occurring in the practice of a celebrated Uroscopian physician concerning the detection of a case of injury to the spine by examining the urine, which runs thus: A gentleman, whose wife had been accidentally precipitated down a flight of stairs, thereby receiving serious injury of the spinal column, was persuaded to carry some of the morning urine of his afflicted wife to the "water-doctor" for examination. Being somewhat incredulous as to the ability of the doctor to determine the case, with an air of coolness he presented the case, saying:

"Here, doctor, is the urine of my wife. I want to *see* if you can tell me what ails her. Now, ask me no questions."

The Doctor finding the urine to be strongly alkaline, and to contain abundance of deposits, of phosphate of lime, readily determined it a "spinal affection," proceeding, most likely, from injury by a fall, wrench, blow, or contusion of some kind, etc., to the silent astonishment of the husband. At length he remarked:

"Doctor, I see you know *all;* but some one has *told* you it."

"Not at all," was the reply, "I know nothing of the case, only from the urine."

"Well, then," thought he, "we will try you a little farther," and he resumed thus:

"Doctor, can you tell me how many steps she fell down? answer *that,* and I will believe."

Here was a poser. But here substituting subtility for science, the Doctor took another peer at the urinal, and *presum-*

ing she might have fallen down the whole flight of stairs, *guessed* the number at *ten*.

"Ah ha! you have missed it! I knew you could n't tell anything," exultingly responded the husband; "why Doctor," said he, "she fell down the whole *thirteen steps*, from top to bottom."

"Indeed! and did you bring *all* the urine she passed this morning?" asked the Doctor, (taking another peer at the urinal).

"No, certainly not?" responded the husband.

"How then do you expect me to tell the whole number of steps down which your wife was precipitated, without bringing it *all* along. You *see*, you left the other *three steps at home*," said the Doctor, in a seemingly angry tone.

"I see, I see, it 's all right, put up the medicine, and tell me your fee"—was the reply.

BLOODY URINE.

WHENEVER the elements of blood appear in the urine, there is ample proof of the existence of active or passive hemorrhage, either in the kidneys, bladder, urethra, or (in the case of a female), from the vagina or uterus.

Dr. Simon has given us some data whereby we may pretty nearly determine from which part the blood in urine may proceed. He says: "Blood flowing from the urethra comes in drops; if the blood is discharged in masses after clear urine, it comes from the bladder, and in that case it often stops up the passage from the bladder by coagulation; if the blood is distributed through the veins partly dissolved, and not in very large quantities, it comes from the kidneys; if it be dark and mixed with pus, it owes its origin to an ulcer." And we add, if it be mixed with mucus it is likely to proceed from the vagina or uterus; and if, as mentioned in speaking of granular kidney, the urine contains albumen, it is from the kidneys also.

Bloody urine may proceed from a laceration of the internal surface of the kidneys, bladder, or urethra, by mechanical violence, in the passage of renal calculi; or it sometimes is occasioned in the course of certain other disorders which affect the system more at large, as scurvy, etc., and at other times is a symptom of unpropitious moment in typhoid fever, and of considerable moment in small-pox, measles, etc., the disease in these cases will always be known by the characteristic combination in the urine. (For microscopic appearance of the blood in urine, see Chart, Fig. 11.)

The microscope has often afforded valuable assistance to the pathologist, not only in a medico-legal point of view in detecting blood from colored fluids in cases of murder, but in distinguishing human, from other blood-corpuscles, in cases of imposition.

"Some years ago," says Smith, "we were summoned to see a dyspensary patient laboring under bronchitis, who was spitting florid blood. On examining the sputa with the microscope we found that the colored blood-corpuscles were those of a bird. On my telling her that she had mixed a bird's blood with the expectoration, her astonishment was unbounded, and she confessed that she had done so for the purpose of imposition."

Treatment: In bloody urine the treatment should be varied according to the part from which the blood proceeds, or the cause which produced it. Spirits of turpentine, balsamic preparations, gallic acid, vegetable astringent durctics, etc., are indicated.

PURULENT URINE.

Pus often appears in the urine of persons laboring under suppurative inflammation of any part of the urinary apparatus, and sometimes is derived from abscesses of adjoining parts, which discharge their contents into the kidneys or bladder. It is said that the purulent contents of a diseased pleura even, have been known to escape into the kidneys and be discharged with the urine.

Dr. G. Bird has given the characters of urine containing pus, in about as concise a manner as can well be done, which amounts to something like this: "Urine generally acid or neutral, unless long kept, and is always slow to assume putrefactive change; by repose, pus falls to the bottom, forming a dense homogeneous layer, of a pale-greenish cream-color, seldom hanging in ropes in the fluid like mucus does, and it becomes, by agitation, uniformly diffused again through it. The addition of acetic acid neither prevents this diffusion, nor dissolves the deposits. When a drop of purulent urine is placed under the microscope, the particles become visible; they are white, roughly granular, exteriorly, and are much more opake than blood-corpuscles. By close attention to the above facts, we will easily diagnose purulent urine, a matter of very considerable importance in many diseases. By agitating with ether, pus gives yellow, butter-like globules of fat; it is never a constituent of healthy urine, and affords diagnostic information of value, not otherwise easily obtained. (For microscopic appearance of pus in urine, see Chart, Fig. 13.)

MUCOUS URINE.

Mucus, in very small quantity, is so generally present in urine, that when there is merely sufficient to form a visible cloud, it is considered by some to be normal. But when it is found to be considerable in quantity, or to obtain in excess in that fluid, an irregular, gelatinous mass, often entangling large air-bubbles, will be observed, which no agitation, however violent, can completely mix with the urine. This, of itself, is generally sufficient to distinguish the character of mucus from that of purulent urine, which it somewhat resembles, and the only kind with which it is likely to be confounded.

This state of the urine is always connected with an irritated or inflamed condition of the genito-urinary mucous membranes. Chronic inflammation of the bladder is a frequent source of this kind of urinary discharge, together with whatever produces inflammation of the urinary organs, as calculi, etc.

To distinguish it from pus by chemical test, add a little acetic acid, which will coagulate it into a thin, semi-opake corrugated membrane. Unlike pus, agitated with ether, mucus never gives but mere traces, if any, of fat.

We have had a number of cases of mucous urine, some of which had been treated by eminent physicians for a long time, without benefit, which were soon relieved by our treatment. If the failure was the consequence of mistaken diagnosis, which we believe, it is but another strong argument in favor of the Uroscopian system. (For microscopic appearance of mucus in urine, see Chart, Fig. 11).

ANIMAL URINE.

Many anecdotes are told, of the attempts at deception, practiced upon Uroscopian physicians, by presenting the urine of animals for inspection. Some persons are so ignorant as to to suppose, and even some physicians have asserted, that it is impossible to tell the urine of a man from that of a horse, cow, etc. This, however, needs no refutation, to those who can lay any claim to medical knowledge, as all medical men know that there is so wide a difference, both in appearance and constituents, that a mere optical glance would be sufficient to detect the attempted imposition, to say nothing of its strong odor, its microscopic appearance, or the reliable chemical tests. Anyone that has ever been far enough from home to pass the barn-yard, need not be reminded of the strong ammoniacal odor arising from the urinary products of the horse or the cow, and this alone should be sufficient olfactory tuition to satisfy one, on the first opening of the vial containing urine collected from these animals, of the absurdity of the attempt at such deception.

The milky color, the specific gravity, loaded as it is with other salts not belonging to human urine, together with the chemical and microscopic examination, will never fail to detect it, even though a portion only of the specimen should be of the animal kind.

Should chemical and microscopic examinations be requisite at any time, the large quantity of *hippuric acid* contained in animal urine, and which is not an ingredient of that of human, will very readily determine the product.

Evaporate a small quantity to the consistence of sirup, then add an excess of hydrochloric acid, when hippuric acid, if any, will fall to the bottom of the vessel in large crystals, and may be examined by the microscope. (See Chart, Fig. 6).

Prof. Leibig gives an account of a girl laboring under hysteria, who refused all food except apples, of which she devoured an immense quantity. Her urine contained a large quantity of hippuric acid; like the urine of a horse or cow.

A neighbor physician, a *curioso*, by the way, being somewhat chagrined at the success of his Uroscopian out-rival, undertook the foolish task of deception by presenting the urine of a horse, requesting an examination of the case, a written medical opinion, and of course, a prescription, if it was thought the patient required it.

The Uroscopian proceeded to examine the specimen and was not long in detecting the imposition. He continued to prosecute the examination—chemically, by a number of analytic tests, and finally by close microscopic scrutiny, in each writing down the results; and lastly, summing up his opinion and prescription thus:

"Urine of an herbiverous quadruped.

"Disease, *Inanition* (emptiness).

"Prescription: *Cerealea*, (sufficient quantity). To be given three times a day.

"Charge: *Twenty Dollars.*"

Slowly, but with determined look upon his customer, he rose to his feet, with pistol in hand, and presented the above prescription. Suffice to say, the bill was liquidated without *demurring*, and neigbor Dr. Curioso, went home a "satisfied" man.

URINE OF PREGNANCY.

We are often called upon to diagnose cases of pregnancy alone, or to decide between this and diseases in which cessation of the menses is involved.

In general, there is but little difficulty to determine a case of pregnancy by an examination of the urine alone, especially if there is no disease in complication. When a female becomes pregnant her health generally improves, whatever it may have been before. The urine, therefore, in this improved condition, will be more natural in appearance and composition, than if there was suppressed menstruation from some other cause. And from whatever other cause such suppression may proceed, the characteristic mark of such cause will be exhibited in the urine, and be as a kind of negative sign.

Suppression of menses with normal urine is a suspicious indication, but by no means a reliable one, unless the well-known principle called *kiestein*, a constituent of the renal secretion of women during utero-gestation, be detected also.

To describe the mode for its detection, although well known to practitioners generally, will not be out of place here.

By allowing the urine of a pregnant woman to repose in a cylindrical vessel, a cotton-like cloud first becomes visible, "which in the lapse of time," says Bird, "varying from the second to the sixth day of exposure, becomes resolved into a number of opake bodies, which rise to the surface, forming a fat-like scum, remaining permanent for three or four days. The urine then becomes turbid, and minute flocculi detach themselves from the crust and sink to the bottom of the vessel; this action continues until the whole pellicle disappears."

It is distinguished from other kinds of pellicle, from its never becoming mouldy, or remaining on the surface more than three or four days.

This investigation, carefully made, is one of the surest tests of pregnancy, and may be relied on with perfect confidence.

Dr. Golding Bird experimented extensively in this way, with the urine of married and unmarried women, and with that of the pregnant and those not pregnant, and places great reliance upon it as a test of that state.

A great number of writers have given attention to this matter, and although some may differ in their views respecting the manner of formation of kiestein, its general presence in the urine of pregnant females is conceded.

Griffith, Reese, Markwick, Bowman, and a number of others, have laid down the same rule as above for its detection. And Dr. Griffith says: "With the exception of the peculiar fermentation which produces the kiestein, and the peculiar cheese-like odor, there are no characteristics of the urine in pregnancy."

Dr. Reese says: "There is a substance of a caseous character, called kiestein and gravidine, found in the urine of women during gestation."

Dr. Markwick says: "This substance (kiestein), is principally met with in the urine of pregnant women, in the form of a greasy, fat-like pellicle, which, after having remained stationary for a few days, breaks up and gradually falls to the bottom of the vessel."

Dr. Bowman says: "The peculiar form of mucilaginous or caseous matter, usually present in the urine of pregnancy, and which has received the name of kiestein, gives the urine containing it a cloudy appearance, and, after the lapse of a few days, gradually forms on the surface in a more or less shining pellicle, which, in three or four days, as the urine becomes ammoniacal, breaks up into small particles and subsides at the bottom, etc."

When such men as the above, after having fully tested the matter, are so uniform in their opinions respecting it, we may be allowed to say, that we have examined a great number of cases in the same way, and can recommend it as giving satisfaction in every case so examined, where another disease has not so far altered the chemical qualities of the urine as to enshroud this substance.

An amusing anecdote is told of a gentlemen residing in the country, who being himself somewhat indisposed for some considerable time, collected a specimen of his morning urine in a vial, for the purpose of obtaining the opinion of a Uroscopian physician upon his case, placing it upon the mantel in his chamber, until he should prepare to start. Breakfast over and his horse brought out, he pocketed the vial and was off. On arriving, he presented the vial to the Doctor, saying: "I want you to examine this case closely, and tell me all about it." The normality of the urine and the cotton-like cloud led the Doctor to suspect pregnancy, and, upon closer scrutiny, to pronounce more positively, that the patient was *enciente*, certainly pregnant. The gentleman gave a derisive laugh. The Doctor persisted in his opinion. The gentleman gave a most contemptuous laugh, deriding the Doctor to scorn, declaring "the urine is my own," etc., whereupon he was *booted* out of the office.

Returning home in a rather unpleasant mood, he there related the above-mentioned circumstances, dwelling especially upon the absurdity of the Doctor's *opinion.* The servant girl, who was at this time an unperceived but anxious listener, was soon observed bathed in tears. Upon inquiring the cause of her tears, she stated that she had emptied the vial on the mantel in the morning, and refilled it with her own urine. This explained all. The sanctity of her virtue had been invaded, and her condition was now fully known. Science had revealed to the Doctor a fact, which time only proved to be "too true" for the girl.

SPERMATORRHŒA.

This disease is generally the result of a most degrading vice, and is one that produces consequences upon the human system, fearfully to be deplored: not only merely because of the *drain* upon the nervous system, and the loss of the vital or nervous fluid in the semen, but also of that long train of nervous symptoms which it is likely to produce upon the body; and the final reduction of the mental faculties which it so much and so soon debilitates.

It is characterized, as its name indicates, by an excessive flow or discharge of the semen; and the animalcules called spermatazoa, are readily detected in the urine of a person afflicted with this disease.

"In some cases, "says Bird, "a sufficient quantity of spermatic fluid is found mixed with the urine, to form a visible cloud, and becomes an important guide to the practitioner in the investigation of a case, perhaps previously and otherwise obscure." The escape of semen is not always the result of bad habit, or the vice of self-pollution, but may arise from extreme constipation, fatigue, paraplegia, excessive use of stimulants, etc. Urine containing spermatic fluid will be found cloudy and opalescent, and will redden litmus-paper. It maintains this cloudy appearance, whether by the application of heat or nitric acid, but the addition of nitric acid often produces a slight troubling in this kind of urine. It is sometimes as opake and thick as if mixed with gruel, and has a peculiar characteristic, fetid and nauseous odor; but no characteristics can be given the urine, making it as distinctly diagnostic of the presence of semen, as may be assumed in the examination by the microscope, although should a large quantity of that secretion be present in the urine, it can generally be recognized by these physical characteristics.

For microscopic examination, let the urine repose in a glass vessel for a short time, decant all the fluid except the last few drops, place one of these on a slip of glass, cover it with a piece of mica and place under the field of the microscope: "The spermatazoa will be observed as minute, ovate bodies, provided with a delicate bristle-like tail, which becomes more distinct on allowing the drop of urine to dry on the glass," says Bird, "and that the detection of spermatazoa in the urine will often enable the physician to discover a source of exhaustion previously concealed from him, and baffling his treatment, is unquestionable."

The discovery of spermatazoa on the person or clothing even of the subject, has been made a test in medico-legal investigations, for the proof of rape. When found on the person of the subject it is conclusive evidence, if not that a *rape* had been committed, at least that emissions of a male had occurred, and would go far as corroborative testimony in the case. (See Chart Fig. 12, C).

Treatment: Abstinence from exciting causes, stimulants, etc. Aperients, cold hip-baths, bromide of potassium, catkins of willow, lupulin and camphor, digitalin, etc., have all been recommended. Cauterizing the seminal ducts, is recommended by Lallemand. Much will depend on the influence over the moral feelings, exerted by the patient himself; without which favorable influence, medicine alone is useless.

TORULA.

When urine contains even a small amount of sugar, too little even to cause it to assume a diabetic character, certain phenomena are developed which will at once point out the presence of that substance.

When this kind of urine is left in a warm place, a scum soon forms on its surface, as if a little flour had been dusted on it. This consists of small oval vessicles containing in their interior, minute granular corpuscles, which expand and become tubular, and finally put on a beaded or jointed appearance. These are colled *torulæ.* (See Chart, Fig. 12). They are only met with in saccharine urine while undergoing alcoholic fermentation. The tests for this kind of urine are more fully given under the head of Diabetes, as in that disease the greatest abundance of saccharine matter is found in the urine. We will, however, here insert the directions for the application of "Moore's test":

"Place in a test-tube, about two drachms of the suspected urine, and add nearly half its bulk of liquor-potassa. Heat the whole over a spirit-lamp and allow actual ebullition to continue for a minute or two; the previously pale urine will become an orange-brown, or even a blister tint, according to the proportion of sugar present." Diseases in which torulæ are found are, diabetis mellitus, with great lassitude and deficient sexual appetite. The urine is generally passed in very large quantities, and is apt to be deficient in urea.

VIBRIONES.

THESE are minute animalcules, which are occasionally met with in the pale, light urine of cachectic and debilitated persons. "All the urine in which we have found these minute creatures," says Bird, "has been pale, neutral and of low specific gravity, and rapidly underwent putrefactive fermentation. When a drop of such urine is examined with the microscope, it will be found full of minute, linear bodies, moving with animation." They have an oscillating motion, "strong enough to excite tolerably rapid currents in the fluid." (See Chart, Fig. 9 C). They are only met with, in the urine of persons in an excessively low and depressed state. In some scrofulous cases, of extreme prostration, in which the powers of life are fast ebbing away, or in greatly debilitated cases of syphilitic disease, where there is extreme prostration of strength, they are more abundantly developed. In fact, in these latter cases, their appearance in the urine may be considered as *the* characteristic of that state.

Alteratives, chalybeates, and stimulant tonics, would seem to be the only course of treatment that would be of service. Sea-bathing and fresh air, where the patient is able to stand the exercise, might be assistants to medication. Some saline diuretics might be of much benefit, "as there can be no doubt," says Dr. Millington, "that the kidneys are glands for draining away that from the blood which would be detrimental to its action if it remained in it, and which is not simply water, but often other matters that lead to disease."

CANCER OF THE WOMB.

Cancer is a vascular, morbid production, characterized by a form of organic cell, which is peculiar, and never enters as a constituent in any normal tissue.

It is a hard tumor, which ultimately terminates in a peculiarly fetid and ichorous ulcer. It requires for its production a peculiarly cachectic condition, and occurs generally in those patients only who may be considered as having a cancerous diathesis. "Very great difficulty has often been experienced in arriving at an accurate opinion as to the cancerous or non-cancerous nature of a tumor, and it is only by carefully noting the collective appearances observed, upon a microscopic examination, that we shall be enabled to decide," says Dr. Beale. But Dr. Donaldson asserts positively that "true cancer can be distinguished from every other tissue, normal or pathological, by certain clear and well-defined elements," among which "its lactescent, infiltrated juice, is very characteristic. The presence of this peculiar fluid is of itself a point of differential diagnosis of great value, the microscope always detecting in it, when found, the presence of cancer-cells," etc.

Anyone at all acquainted with the sanious ichor secreted from a cancerous surface, and the peculiar odor arising therefrom, would not be long, at least, in suspecting it, from the very offensive smell itself.

"The odor which attaches itself to it, is alone sufficient, in forty-nine cases out of fifty to establish a diagnosis," says Bennet. "More especially is this the case in the horribly offensive odor of a cancerous *uterine* discharge. It is so peculiarly nauseating, as to leave a lengthened impression on the olfactory nerves, the least portion of which excreted in the urine will be readily discovered."

In cancer of the uterus, the microscopic examination of the discharges becomes highly important in arriving at an accurate diagnosis. "Cancer-cells, in such cases, may often be detected in the discharges," says Carpenter. The microscopic appearance of these cells from the uterus is represented in Chart, Fig. 15 C, as they are observed in the urine of such patients.

The caudated cancer-cells, as represented in Chart, Fig. 16 A, are of common occurrence in cancerous tumors, and in cancer of the bladder are invariably present. Therefore they should be carefully examined. The concentric cancer-cells, as represented in Chart, Fig. 16 B, are those of cancer of the breast, ovaria, etc., and deserve to be closely scrutinized and well understood, in order to form a correct diagnosis. It is in uterine cancer that the microscopic inspection of the urine overrides in value all the other means for correct diagnosis.

LEUCORRHŒA—(Whites).

This disease affects the mucous membrane of the vagina and uterus, the principal local symptoms of which are the pouring out of an unhealthy secretion, or an unnatural, discharge, of a whitish, opake, or sometimes, yellowish or even greenish color. Most writers include this latter color under gonorrhœa, but this is highly improper, where the true gonorrhœal virus does not exist, as the pathological conditions of the parts in the two diseases are entirely different. There is frequently a great similarity between them, however, and much mischief and injury may be occasioned by want of proper discrimination.

The leucorrhœal discharge is more like jelly and is generally whiter than in gonorrhœa, and comes, as before remarked, from the vagina and uterus, while the seat of gonorrhœa is more generally in the urethra, and the discharge of a more specific, contagious character.

Uterine and vaginal discharges present somewhat different characters in different specimens. Mucus, pus, blood, and epithelium, from leucorrhœa, is represented from a microscopic examination, in the Chart, Fig. 14.

"In leucorrhœa many of the epithelial cells are filled with oil-granules, and mingled to a greater or less extent with pus-corpuscles."

By microscopic examination of a drop of the urine containing it, the leucorrhœal discharge may be more readily distinguished from gonorrhœal than by any other means.

Vaginal epithelium is always of the scaly variety, and consists of large, flat, ragged, and very irregular cells, folded over each other, and perhaps creased in different directions.

GONORRHŒA—(Clap).

This disease is characterized by a discharge from the mucous membrane of the urethra, produced by contagion from a similar discharge during sexual connection.

This matter is generally, or always, of a yellowish or greenish color, and is of a highly contagious character. Although this virulent poison may produce gonorrhœa on the mucous surfaces of the body, as the eyes, nose, etc., it is pretty generally confined to the urethra, producing a considerable degree of pain and scalding heat, in passing water. In addition to the gonorrœal matter which is discharged from the urethra, there is sometimes a tinge of blood in the urine also.

Although this disease may generally be quite correctly diagnosed in the male, subject, a great difficulty is sometimes experienced in distinguishing it from leucorrhœa, in females. Great injury might be done, both to the character and constitution of the patient, by a mistake in the diagnosis. Therefore great caution is required of the physician, lest by a mistake in the diagnosis, he might cast reflections on the continence of a chaste woman, and hazard his reputation and property, by incurring a suit for mal-practice.

"That medico-legal science has been greatly enriched and rendered far more certain in its results by the aid of the microscope, few persons will deny," says Carpenter. "The ends of justice sometimes depended solely upon its power of detecting spermatazoa, in the case of rape, of distinguishing between the stains of blood and those of colored fluids, or of pointing out the difference between human hair, and that of animals," etc.

And in the microscopic examination of the discharge in this disease, the most convincing proofs are obtained. "The peculiar appearance of the epithelial cells," says the above author, "will indicate the part of the genito-urinary mucous membrane from which the mucus or pus was secreted." (See Chart, Fig. 11, F).

The peculiar yellowish or greenish color of the purulent matter in gonorrhœa, together with the above-mentioned epithelium, will enable anyone to detect and discriminate between this and the preceding disease.

Urine mixed with the matter of gonorrhœa is always acid to test-paper. An alkaline wash to the parts, with alkaline diuretic salts would therefore be appropriate treatment, in connection with medicines to destroy the specific, contagious virus.

Balsam of copaiba has long been in great use, but is not alone to be relied on. Podophyllin, irisin and phytolaccin, in connection with the alkaline treatment, will be good.

SYPHILIS.

THIS disease is always produced by a specific poison, the smallest particle of which is sufficient to bring on the disease in all its virulence.

This poison is imbibed by the blood and transmitted through the stream of life, until sooner or later, the entire mass becomes affected, and the whole body contaminated with the disease.

"It is a well-known law of animal poisons," says Wilson, "that being once introduced into the blood, they excite in that fluid an action which has for its object the production of a similar poison, and this process goes on until the blood becomes saturated or overcharged with the morbific principle."

The poison of syphilis having once entered the system, it is hard, if even possible, to ever thoroughly eradicate it. At all events, it remains for years, and sometimes for life, giving notice of its existence from time to time, perhaps, by a variety of symptoms, more especially if an appropriate and thorough course of medication has not been promptly meted out at the onset almost of the disease.

The urinary secretion will early give evidence of this poisonous matter, as this is one of the great emunctories whereby poisons are eliminated from the animal economy.

In a short period of time, the whole system becomes disturbed by this innoxious principle of the blood, and all the organs and structures supplied with that blood must suffer, to a greater or less extent, and the disease becomes a constitutional one.

In this peculiar cachectic state of the system, the urine is pale, but not clear, neutral in its character, of low specific gravity, and very rapidly undergoes putrefactive fermentation.

When excessively debilitated from this cause, a drop of urine examined under the microscope, will exhibit a great number of minute, linear bodies, called "vibriones." (See Chart, Fig. 9, C). "These have an oscillating motion," says Bird, "strong enough to excite tolerably rapid currents in the fluid."

This is essentially *the* characteristic of syphilitic urine, and may be seen in the urine of even less-developed cases of syphilis, but is always to be found in that of the extremely debilitated, or prostrated from that cause.

Treatment: Alteratives, chalybeates tonics, and stimulating diuretics, are indicated in all such cases.

HEADACHE.

This frequent and distressing disease, although apparently arising from different causes, yet seems to be dependent upon some peculiar condition or action of the blood. A marked derangement of the stomach produces its effect upon the circulatory system, a recession from extremities or some other part, and a determination of that fluid to the head follows.

"In a majority of cases where persons are subject to frequent returns of headache," says Prof. Newton, "we are satisfied that the cause is deficient secretion of the solids of the urine."

We have noticed in such cases that the urine was secreted in small quantity, and was of low specific gravity previous to, and at the time of the attack.

Such headache becomes more or less periodic, frequently, and requires to be treated with antiperiodics, in connection with diuretics, podophyllin, leptandrin, valerianate of quinine and iron, with acetate of potassa, etc.

Dr. Simon says: "Now, it is a common observation that almost all headaches are most promptly relieved by whatever stimulates the kidneys to throw off a quantity of urine. When this takes place the system feels at once relieved of a load or oppression which seems to clog all its energies, and the mind as well as the body, becomes more light and vivacious."

With the improvement in the urinary secretion will the relief of the disease be manifested.

DENTITION.

The following remarks on the above subject, by Dr. Simon, we have thought worthy of a place here. Dr. Simon prefaces them by alluding to the frequent inactivity, and sometimes the almost complete suspension of the functions of the kidneys, etc., while the system of the child is suffering severely from dentition.

"The simple question as to the quantity and color of the urine (and by the way, we can much better trust the report of nurses about the state of the urine than we can about the alvine evacuations), will often enable us to at once form a correct opinion as to the general or constitutional health of our patient. As long as the kidneys act freely, there is little or no risk in the symptoms of mere dentition, however severe and distressing the signs may be.

"The same remark is, we believe, strictly applicable to the prognosis of most cerebral affections in children. When the urinary secretion is scanty and deep-colored, the circulation seems to be oppressed and excited; and the rapid, on some occasions, almost instantaneous mitigation of the alarming symptoms, after a copious discharge of water, is well known to all experienced practitioners.

"On the whole, we do not think that there is a more important sign to be attended to in the management of children, during the first two years of their life, than that afforded by the state of the urinary secretion.

"In fine, the kidneys become, in numerous cases of disease, the seat of an active, eliminatory process, of which the skillful physician will avail himself in the treatment of dentition, and of various other affections, to which children are especially liable during the first two years of their lives."

We have given the above, both for the practical remarks it contains in favor of the urino-pathology, or the "Uroscopian System," as well as the value of such investigations in the treatment of dentition.

If it be true, that almost all headaches are promptly relieved by a free discharge of urine, as observed in the article on that subject, how important the examination of that discharge must be in all cerebral affections, and how derelict in duty would be the physician, who, through ignorance or neglect, would not avail himself of the knowledge necessary for the prompt relief of these little sufferers.

If the irritation of dentition, however, is as yet confined to the stomach, the urine will usually be pale and abundant, depositing a whitish or yellowish-white sediment.

WORMS.

THE troublesome complaints among children, occasioned by worms, are many times difficult to distinguish from other diseases by the ordinary mode, and frequent mistakes occur in the treatment of children, for want of a proper diagnosis, the little patients being harassed with a long course of medication for infantile remittent fever, or some other disease, when worms only were the first cause of illness; or drenched with nauseating "vermifuges," when the patient was laboring under disease of a more formidable character, and in which these medicines were productive of the worst consequences.

Every intelligent mother, however, who has had any experience in rearing children, is enabled to form some idea of the state and appearance of the urine in these cases, and will describe it as being sometimes thick and yellowish, or more generally of a milky-white color. So almost universally is this appearance of the urine observable in a case of worms, that upon this alone the little patient is many times very correctly hurried off to the physician, before any other symptom is discoverable, when, if properly treated, the more serious consequences of convulsions, etc., are averted.

How important, then, that the physician himself should be enabled to distinguish these cases, and not leave it alone to the nurse to decide.

Professor Eberle, however, says: "In worms the urine is turbid, yellowish—or, after depositing a sediment, has the appearance of milk and water."

Professor Beach says: "The urine is pale, thin, crude, and in some instances the color of whey, or quite white."

Prof. Payne, says: "There are frequent changes in the appearance of the urine, which is at one time thin, scanty and of a milky appearance, and at another copious and limpid."

A little attention to the urine in worms will enable anyone to distinguish it almost at first sight, by its pale, light, milky-white, yellowish tinge, or whitish color. And every physician should acquaint himself with the characteristics thereof, that intelligent and appropriate remedies may be administered. This appearance of the urine, however, is met with only in cases of children troubled with lumbricodes, or tænia solium.

We have in our possession several large pieces of tape-worm which were expelled by our treatment, the diagnoses having been made out by an examination of the urine alone—one of the patients being an adult, had been treated for nearly every disease but worms, before coming under our professional care. But with this characteristic of the urine you need never mistake a caso of these kind of worms.

There are other kinds of worms, however, that infest the human body; and that are to be discovered in the urine, or by their marks in that fluid. The *strongylus gygas*, sometimes occupies the human kidneys, and causes great suffering. We observed one of these parasites that had been expelled from a patient of Prof. E. G. Dalton, of the Eclectic Medical College of Philadelphia. It was several inches in length, and was passed from the urethra. The *dactylius aculeatus*, was first described by Curling—who discovered several of them in the urine of a little girl.

A patient of M. Lawrence, voided numbers of the parasites, *diplosma crenata*, for a length of time, from the urinary bladder.

Also, the *spiroptera hominis*, was discovered in the urine, by Mr. Lawrence.

The most beneficial or generally successful treatment for the removal of lumbricodes will be found in the following:

Pulverized Santonin twelve, Pulverized Podophyllin two

grains. Divide into twelve parts, and give one morning and evening to a child two years of age.

For tape-worm, pumpkin-seed, orgeat and spirits of turpentine, has succeeded best in our hands.

The other kind will require a more diuretic medicine.

DYSURIA—(Of Infants).

This complaint is very frequent among infants, and is oftentimes overlooked or passes unnoticed, because of the inability of the little sufferer to make known the seat or location of the excruciating pain, which is manifested only by shrieks.

This painful affection in voiding urine, is always attended with an unnatural condition of the urinary secretion, and humanity alone should prompt the physician to use every means to discover the true cause of suffering, more especially where innocence is the sufferer, and infancy prevents the communication of the facts.

"In a majority of these cases the urine contains a large portion of lithic acid," says Eberle, "and, occasionally, it is highly charged with phosphoric, sedimentous matter. These substances impart a peculiarly irritating quality to the urine, and, when they are copious, and the system is in an irritable condition, may readily produce a considerable degree of irritation about the neck of the bladder, and give rise to pain and difficulty in passing urine."

When a child becomes affected with pain and difficulty in passing urine, this secretion ought to be carefully examined, both in a recent state, and after it has stood for some time. If the sedimentous matter of the urine be of a red, or reddish color, remedies calculated to counteract the secretion of lithic acid by the kidneys, will be indicated, and will probably procure speedy relief.

Dr. Prout observes that children, in general, and especially those of dyspeptic or gouty parents, or who inherit a tendency to urinary diseases, are exceedingly liable to lithic-acid deposits in the urine. If the urine be examined it will always be found to be very unnatural, and frequently loaded with lithic acid; and should this prove to be the fact, the case requires immediate attention, as there is much greater risk, at this period of life, than at any other, for the formation of stone in the bladder.

In this condition of the urinary secretion, the administration of small doses of magnesia, bicarbonate of soda, subcarbonate of potash, lime-water, etc., would be peculiarly suitable medicines, in connection with some mucilaginous drinks, such as flaxseed tea, althæa officinalis, etc.

The common Neutralizing Cordial of Dr. Beach's work, is very valuable in these cases, because of its alkaline properties, and peculiarly acceptable stomachic in all diseases where acid in the alimentary canal predominates—children who are much affected with acidity in the *prima viæ*, being most apt to experience urinary difficulties of this kind.

ENEURESIS.

THIS disease may be occasioned by the same faulty state of the urine, as in dysuria, and in fact there is many times no real or apparent difference, farther than the mere external symptom of discharging the urine during sleep at night in the one, and the greater degree of pain in voiding it, in the other.

Prof. Eberle says: "Incontinence of urine very frequently, perhaps always, in the first instance, is excited by an acrid condition of the urinary secretion itself. In those cases, especially where the discharge takes place in consequence of involuntary effort excited by a lively dream, the urine almost always contains an excess of sedimentous matter, particularly lithic acid and its compounds, imparting to it an irritating character."

"Hence," says Dr. Prout, "we have been led to infer that, in this species of urinary incontinence, the acrid properties of the urine are chiefly at fault." He also thinks that some peculiar morbid condition of the urinary organs constitutes the most frequent cause of those cases of nocturnal incontinence of urine, in which the discharge takes place involuntarily, and without any consciousness of its occurrence also.

"From what has been said above," says Eberle, "it need scarcely be observed that, in prescribing for a case of this kind, particularly when of a recent character, the urine ought to be carefully inspected, as a preliminary step in the adoption of a suitable plan of management. Should the urine be found to contain much sedimentous matter, remedies ought to be employed for correcting the urinary secretion. If the lithic-acid deposits predominate, small doses of magnesia, lime-water, the subcarbonate of potash, or of the bicarbonate of soda, should be resorted to," etc. And in cases attended with phosphoric urinary deposits, remedies calculated to counteract the formation of these should be employed, such as vegetable acids, acescent articles of nourishment, etc. By such a course of management, recent cases especially may be completely arrested.

" Incontinence of urine in children sometimes depends upon a morbidly irritable state of the bladder. The patient during the day is more or less harassed with a frequent desire to urinate, and the discharge is always accompanied with considerable uneasiness, and sometimes with much pain. These cases are usually associated with a morbid condition of the urine, also ; sometimes with an excess of lithic acid, and occasionally, with phosphoric deposits. Instances of this kind must be managed in the way stated above."

INDEX.

www.ingramcontent.com/pod-product-compliance
Lightning Source LLC
Chambersburg PA
CBHW021519210326
41599CB00012B/1307

* 9 7 8 3 3 3 7 7 1 5 0 9 0 *